g is continuing to increase rapidly across multiple disciplines. This
ies on International Perspectives on Aging provides readers with
mprehensive texts and critical perspectives on the latest research,
tical developments. Both aging and globalization have become a
nes, yet a systematic effort of a global magnitude to address aging is
he series bridges the gaps in the literature and provides cutting-edge
nd traditional areas of comparative aging, all from an international
ore specifically, this book series on International Perspectives on
spotlight on international and comparative studies of aging.


ion about this series at http://www.springer.com/series/8818

# International Perspe

## Volume 27

**Series Editors**

Jason L. Powell, Department of Social ¿
Chester, UK
Sheying Chen, Department of Public Ad
NY, USA

The study of agin
wide-ranging se
much-needed co
policy, and prac
reality of our tir
yet to be seen. T
debate on new
perspective. M
Aging puts the

More informa

Philip D. Sloane • Sheryl Zimmerman
Johanna Silbersack

Editors

# Retirement Migration from the U.S. to Latin American Colonial Cities

 Springer

*Editors*
Philip D. Sloane
University of North Carolina at Chapel Hill
Chapel Hill, NC, USA

Sheryl Zimmerman
University of North Carolina at Chapel Hill
Chapel Hill, NC, USA

Johanna Silbersack
University of North Carolina at Chapel Hill
Chapel Hill, NC, USA

ISSN 2197-5841  ISSN 2197-585X  (electronic)
International Perspectives on Aging
ISBN 978-3-030-33542-7  ISBN 978-3-030-33543-4  (eBook)
https://doi.org/10.1007/978-3-030-33543-4

© Springer Nature Switzerland AG 2020
This work is subject to copyright. All rights are reserved by the Publisher, whether the whole or part of the material is concerned, specifically the rights of translation, reprinting, reuse of illustrations, recitation, broadcasting, reproduction on microfilms or in any other physical way, and transmission or information storage and retrieval, electronic adaptation, computer software, or by similar or dissimilar methodology now known or hereafter developed.
The use of general descriptive names, registered names, trademarks, service marks, etc. in this publication does not imply, even in the absence of a specific statement, that such names are exempt from the relevant protective laws and regulations and therefore free for general use.
The publisher, the authors, and the editors are safe to assume that the advice and information in this book are believed to be true and accurate at the date of publication. Neither the publisher nor the authors or the editors give a warranty, expressed or implied, with respect to the material contained herein or for any errors or omissions that may have been made. The publisher remains neutral with regard to jurisdictional claims in published maps and institutional affiliations.

This Springer imprint is published by the registered company Springer Nature Switzerland AG.
The registered company address is: Gewerbestrasse 11, 6330 Cham, Switzerland

# Foreword

At one extreme, cities can be seen as hubs of innovation and progress, full of energy, culture, and life. At another extreme, they can be seen as noisy, depressing, artificial, and crowded yet isolating. From either point of view, the trend toward human urbanization is unmistakable. It is global and historic, and it may also be our planet's future – environmentally, economically, and culturally.

In April of 2019, *National Geographic* published a special issue titled "Cities: Ideas for a Brighter Future" in which the central concept held that sustainable cities were not necessarily bad for the environment; in fact, they could be ideal solutions for many of the challenges faced by our growing and aging populations. The magazine issue was born in part from our internal conversations on the subject of cities and from a request for grant proposals (RFP) from researchers around the world for projects that would address, in new and unique ways, how the future of our planet would be one in balance with nature, yet still urban, globally interconnected, and cosmopolitan.

This book by Dr. Philip Sloane and colleagues, spurred in part by a *National Geographic* RFP Grant, represents an ambitious and comprehensive look at a unique challenge faced by cities in many developing parts of the world: how do cities with deep cultural roots accommodate to international retirement migration from very culturally distinct, often wealthier, parts of the world. In his *National Geographic*-funded project, Dr. Sloane and colleagues studied two Latin-American cities, Cuenca in Ecuador and San Miguel de Allende in Mexico, both of which have seen high migration from American retirees. The team employed an array of research methods and sources, including interviews with local residents, online questionnaires, geospatial mapping, social media, local media, field notes, and photography. In doing so, they covered a breadth of topics ranging from the changing quality of life for a retired migrant to how the exodus was changing healthcare access across nations and impacting local languages and traditions.

In the fourth chapter, for example, the team looked at the difference between the two cities in how local residents' attitudes were changing based on how long retirees had been there. San Miguel de Allende, in Mexico, has seen migrant retirees for decades. The population of Americans, or "gringos," is about 10%, having been arriving there since the 1940s. "Seventy percent or so of the Americans have integrated," explained a native resident. They are a part of that modern Mexican city. In contrast, in Cuenca, the American retiree population is less than 2%, and the economic and social changes that arrive with that influx have not been felt for as long, "I think it is good that they move here," a Cuenca resident said, "but they should accept our rules and procedures, and they should treat Ecuadorians as they want to be treated."

This fascinating work is one particular view of how cities around the world are changing. It provides a glimpse at how cities will continue to evolve in the future, especially as international air travel grows, social media continues to bridge great distances from the convenience of a cell phone, and our vast planet grows ever smaller and more crowded. Social scientists often paint a picture of doom as the world population creeps toward an 11th digit (10 billion by 2050). However, this study is a great example of how global interconnectedness has strong merits, how cities and cultures within those cities are malleable and continue to change and evolve, and how retiree migration can be a win-win scenario for those arriving and for those already there. The authors show us a light at the end of that airport runway.

Anthropologist and Senior Program Officer                            Miguel Vilar
National Geographic Society
Washington, DC, USA

# Acknowledgments

We would like to thank our local coordinators from each city, who assisted with the logistics, arranging of interviews, and shaping of the research over the development of the research protocol and throughout data collection. Without Kati Alvarez, a sociologist based at the Universidad Central of Quito, and Roberto Robles Iniestra, a retired engineer who lives in San Miguel de Allende, this work would not have been possible. Although we were not able to visit Granada, Nicaragua, at the time of this research as originally intended, we would like to thank Milton López Norori, MD, MPH, based at the Universidad Nacional Autónoma de Nicaragua, as well, for his input on our research methods.

We are grateful to our team of data collectors, Luisa Beatriz Pereira Cesar, MSW; Karla Vanesa Jimenez-Magdaleno, MPH, MCRP; Brenna Jean McColl, MSW; and Erika Sheryn Munshi, from the University of North Carolina at Chapel Hill's School of Social Work, Gillings School of Global Public Health, and College of Arts and Sciences. Each of these researchers dedicated themselves to conducting on-site local interviews and networking, assisting with initial analysis and coding of the data, and giving thoughtful advice and input throughout the project.

Analysis of the data collected over the course of this project was carried out by a team of coders, Lea Efird, MSW; Yasmin Barrios, PhD, MPH; and Ailet Reyes, who reviewed and coded hundreds of pages of transcripts in Spanish and assisted with the translation of quotes to English. We are appreciative of the hours and work they dedicated to analyzing this research. Additional help with quote translation was provided by Martha Maria Moreno Linares.

We would also like to thank and recognize the communities of Cuenca, Ecuador, and San Miguel de Allende, Mexico, for their participation in this research. It is through the combined efforts and participation of our local interviewees and the retired expatriate respondents who met with us and responded to our online surveys that we are able to publish these findings. Out of respect for confidentiality, we are not naming individuals, but we could not have been able to gather these data without the guidance and assistance of many members of both the retiree and local communities, and so we cannot overstate our thanks for the welcome we received. We can only hope that this report is helpful and accurate.

Funding was provided by Research Grant # HJ-131R-17 from the National Geographic Society. Additional support was provided by the Elizabeth and Oscar Goodwin Endowment of the Department of Family Medicine at the University of North Carolina at Chapel Hill.

# Introduction

When friends and colleagues talk about retirement, the desire they most commonly express is "I want to travel." This is not surprising – there's something about visiting other places that's exciting and invigorating. Gerontologic research indicates, however, that the most common actual use of time in retirement is not travel but watching television. This reflects the reality that travel tends to be expensive, whereas retirement generally means reduced income.

Imagine, then, something that would accommodate both of these factors – desire for the amenities of travel and the need to live on a budget – and one can instantly see the appeal of international retirement. Indeed, the "pull" factors associated with travel and the "push" factors associated with income constraints are, as the research in this book will explain, two major factors in decisions to retire abroad.

On top of that, international retirement is a media darling. Google "international retirement," and among the 307,000,000 results you obtain,[1] the most prominent will discuss nothing but the appeal, benefits, romance, and affordability of packing up and moving to another country. Anyone with even the slightest itch to retire abroad will be able to read glowing testimonials, view enthusiastic media reports, and receive free materials explaining why retiring abroad is so very much better than languishing at home in front of a television.

But what is international retirement *really* like, and what is its impact on the host communities? I first wondered about these issues over a decade ago while staying with a family in San Miguel de Allende as part of a Spanish study abroad program. One night over dinner, my hosts lamented their cousin's decision to sell the ancestral family home near the city center and move to the outskirts. Retired Americans, they explained, were taking over the central city, driving up prices, and disrupting communities. In response, many locals were selling and moving to newer, less expensive neighborhoods in the "afueras" (outskirts). During my stay in the city, I observed potential upsides to migration as well – from bustling shops to a wide variety of public service projects carried out by the retired Americans.

A few years later, some colleagues and I had the opportunity to study health care among expat retirees from the United States who were living in Panama and Mexico. We found that they struggled with many issues and that their lives and experiences were far from idyllic. Furthermore, as we delved into the literature on retirement to various Latin American countries, we found that academics were often derisive of the retirees, using phrases such as "a contemporary continuation of settler forms of colonialism."[2] Thus, it became increasingly clear that international retirement is a complex issue. Furthermore, most of the research dealt with the immigrant retirees, and little was said about their impact on the local communities – especially in Latin America, where retirement migration is more recent than in Southern Europe.

To address these questions, I applied for and was fortunate to receive a research grant from the US National Geographic Society to study the impact of international retirement migration, largely from the United States but also from Canada and Western Europe, on two cities in Latin America – Cuenca, Ecuador, and San Miguel de Allende, Mexico. I had chosen these cities because, in addition to being popular international retirement destinations, they had strong, deep cultural heritages that would make them susceptible to negative impact from an "invasion" of retirees from the north. Both, like many other Latin American cities, were quite old, with historic "colonial" central areas that had been declared UNESCO World Heritage sites. How, I wondered, would an influx of retirees impact these cities?

To conduct the field research, I assembled a talented, bilingual research team of four students and recent graduates from the University of North Carolina at Chapel Hill (UNC). After reading in depth the literature; consulting with UNC faculty experts in sociology, anthropology, and business; and networking online with the local communities, we collected data during the summer of 2018. The bulk of our time was spent interviewing local residents, as our primary focus was on their perspectives. In order to sample a cross section of the local population, we sought to interview at least six representatives from each of six categories: government officials, health-care providers, non-governmental service providers, real estate agents, and convenience store employees from retiree-dense and retiree-absent neighborhoods. Additionally, we networked extensively with immigrant retirees, conducted Internet-based surveys, and formally interviewed persons who used health-care and long-term care resources. For details on our research methods, please see Appendix.

This book describes the results of that research. Each chapter focuses on a different element of retirement migration; the themes were chosen due to their importance in international retirement migration.

Chapter 1 provides an overview of international retirement migration on a worldwide scale and, in addition, provides background on our two study cities. The chapter reviews the literature in the subject, which until recently has focused largely on the retired immigrants themselves. It presents a theoretical model of the evolution of a city as a retirement destination and San Miguel de Allende and Cuenca as being at different stages of that development. Finally, it introduces the theme most central to the book: the impact of retirement migration on the social and economic structure on the receiving communities.

Chapter 2 presents the original data on the retired expats themselves, placing it in the context of the existing literature. Distinct differences between the retirees in

San Miguel de Allende and Cuenca are apparent, including their age distribution, motivations for moving, and housing preferences and practices. A novel addition to the literature is retiree opinions regarding the positive and negative impacts of retirement migration on the communities they have joined. These findings review a number of their primary concerns, including real estate, which is reviewed extensively in the subsequent chapter.

Chapter 3 presents and discusses data on real estate in the two study cities, focusing on recent trends and perceptions of retiree influence. Drawing largely from interviews with local residents, including real estate agents, it also includes statistical data and results from our surveys of expat retirees to provide a comprehensive picture of this vital economic area and to contrast the two cities. The chapter explores the unintended consequences and changes that come about because of the retirees' greater purchasing power, such as increases in property values and gentrification. It also discusses the potential influences of other factors, such as tourists, second home owners from elsewhere within the country, and nationals living abroad.

Chapter 4 presents what is perhaps the most central aspect of our research – the ways in which retiree immigrants have altered the social and cultural fabric of our two study cities. Here we rely almost exclusively on our interviews with local residents, discussing such areas as retiree integration into society, communication and socialization between natives and immigrants, and the types of relationships that occur between the two populations. The viewpoint of the immigrant retirees is also presented, and all is put into the context of the existing academic literature.

Chapter 5 delves further into the opinions of local residents regarding the impact of retirement migration by focusing on the responses of the 12 local convenience store ("tienda") employees and owners we interviewed in each city. Our research team particularly valued convenience store employee perspectives due to their relative lack of vested interests in the subject (as opposed to, e.g., government employees or real estate agents) and their ongoing connectedness with all manner of community members. In addition to providing an overall perspective on retiree migration, these interviews offer a window into the changing economies of our two study cities.

Chapters 6 and 7 focus on health care – an element that is of particular interest to retirees and has been especially ignored in most research on retirement migration. As two of our team (PDS and SZ) are experts on health care for aging populations, we were particularly well suited to delve in depth into this subject. Chapter 6 focuses on health care overall, including such issues as where retirees go, what they do for insurance, and what they think of health care in the two cities. Chapter 7 focuses on long-term care resources, which most older persons eventually need, but few plan for, and which are only beginning to be developed in our two study cities. These topics will become increasingly relevant as retiree populations age in place.

Our final chapter (Chapter 8) draws from our interview and survey data, as well as the extant literature, in an attempt to present and discuss recommendations for colonial cities in Latin America regarding how to navigate the twin challenges of attracting retirees while minimizing adverse effects on their own population.

As with any large-scale academic endeavor, this book only came to pass through the efforts of many, many individuals, the most important of whom have been named in our Acknowledgments section. It is my sincere hope that the material presented herein is of some value to persons thinking of retiring abroad, those who are currently living abroad, and, most importantly, individuals in the communities that now or in the future will serve as international retirement destinations.

Chapel Hill, NC, USA                                                        Philip Sloane
July, 2019

## References

"International Retirement" Googled on July 27, 2019.

Hayes, M. (2015). Moving South: The economic motives and structural context of North America's emigrants in Cuenca, Ecuador. *Mobilities, 10*(2):267–284.

# Contents

# Contents

# About the Contributors

**Maria Gabriela Castro** MD, is an Assistant Professor of Family Medicine at the University of North Carolina at Chapel Hill where she serves as Program Director for the Family Medicine Residency's Underserved Track at Siler City, NC. Her previous work focused on global health experiences in medical education and on the ethical and educational implications of immersion experiences in resource-constrained settings. Dr. Castro's areas of interest include community health, health equity, and care for disinvested populations.

**Lea Efird** MSW and MPA candidate, is a graduate student at the University of North Carolina at Chapel Hill. She received a bachelor's degree from UNC-CH in 2016, with a double major in Hispanic Literature and Culture and History, as well as a minor in English. Within her Master of Social Work and Master in Public Administration programs, Lea's areas of focus include the US Latinx community, immigration policy, and the integration of social work into the broader public sector.

**Karla Jimenez-Magdaleno** MPH, MCRP, is a Regional Director for the American Voices Project led by Stanford University. Ms. Jimenez-Magdaleno received her dual master's degrees with focuses on Health Behavior and Land Use and Environmental Planning in 2016 from the University of North Carolina at Chapel Hill. Her research focuses on exploring the impact transient populations have on local communities and their physical environment. Much of her approach uses qualitative research and multimedia, including photography, film, and podcast. She has worked on projects about eviction and health, migration post-natural disasters, and economic mobility and development, among others.

**Erika Munshi** BS, is a second-year master's student at Duke University's Nicholas School of the Environment. Her research background includes international work within food security, natural resource management, and global value chains. She received her bachelor's degree from the University of North Carolina at Chapel Hill in 2018, with focuses on Environmental Studies, Global Studies, and Hispanic Studies. She is currently working as an intern at the World Trade Organization on topics relating to trade and environment with a particular focus on climate change.

**Johanna Silbersack** MSW, is a research specialist at the Cecil G. Sheps Center for Aging, within the Program on Aging, Disability, and Long-Term Care at the University of North Carolina at Chapel Hill. She received a bachelor's degree in psychology in 2014 and later received a master's degree in social work with a focus on contemporary social issues. Her work with the Program spans across various issues of aging, including cognitive impairment and the well-being of long-term care staff.

**Philip D. Sloane** MD, MPH, is the Elizabeth and Oscar Goodwin Distinguished Professor of Family Medicine and Co-director of the Program on Aging, Disability, and Long-Term Care at the University of North Carolina at Chapel Hill. A geriatrician and health services researcher, Dr. Sloane has published 19 books, over 200 refereed articles, and over 75 book chapters or other publications. Previous related research includes studies of health-care issues among retirees in Panama and Mexico and numerous studies of long-term care for older persons in the United States.

**Sheryl Zimmerman** PhD, is the Kenan Distinguished Professor of Social Work, an Adjunct Professor or public health, and Co-director of the Program on Aging, Disability, and Long-Term Care at the University of North Carolina at Chapel Hill. A social gerontologist with broad content and methodological expertise, she has published over 300 refereed articles and led over 25 funded studies related to evaluating and improving health services for older persons.

# Chapter 1
# International Retirement Migration

Philip D. Sloane and Johanna Silbersack

International retirement migration from northern, wealthier countries to more southerly countries is a growing phenomenon. It represents one of the many manifestations of the worldwide interconnectedness associated with globalization. In addition to increasing globalization, a number of other key factors play a part in making this movement more popular. In North America, these factors include:

- growing numbers of retirees in good health and with retirement pensions, due to population aging and widespread retirement savings programs;
- increasingly available air travel, which helps many retirees gain familiarity with other countries through tourism, and which reduces the feeling of isolation associated with moving abroad;
- economic globalization, in which the widespread availability of transnational retailers such as Walmart, 7-Eleven, McDonalds, and Target has made living elsewhere less of a "foreign" experience; and
- modern information technology, such as *WhatsApp* and *Skype*, which have made communication with family and friends at home easier and less expensive than in the past.

Furthermore, as cost of living rises in the United States, Canada, and Western Europe, retirees with pensions and savings that must continue to be stretched to support a modest lifestyle at home may meet the resource requirements for immigration to many Latin American countries, where the cost of living is lower.

The first large scale migrations occurred in Europe in the latter half of the twentieth century, as integration within the European Community facilitated ease of population movement and provided reciprocity of health benefits. At the same time, United States and Canadian citizens began retiring in significant numbers to

P. D. Sloane (✉) · J. Silbersack
University of North Carolina at Chapel Hill, Chapel Hill, NC, USA
e-mail: philip_sloane@med.unc.edu

© Springer Nature Switzerland AG 2020
P. D. Sloane et al. (eds.), *Retirement Migration from the U.S. to Latin American Colonial Cities*, International Perspectives on Aging 27,
https://doi.org/10.1007/978-3-030-33543-4_1

destinations in Mexico, both along the coast and in a few interior cities. More recently, international retirement migration has become a truly global phenomenon, increasing rapidly in quantity and in diversity of destinations. While not yet large in scale – far fewer than 1% of retirees relocate internationally in retirement – with the aging of the baby boomers, international retirement is expected to grow markedly (Rojas et al. 2014; Schafran and Monkkonen 2011).

Research on international retirement migration began in the 1990s and has rapidly increased in the diversity of sites studied and methods used (Warnes 2009). The phenomenon has attracted scholars from a variety of disciplines, such as geography, sociology, anthropology, leisure and tourism studies, and economics. It has been variously described as international retirement migration, expatriate retirement, lifestyle migration, amenity migration, and residential tourism.

## 1.1 The Range of Destinations for International Migration in Retirement

International retirement has become a truly international phenomenon. Common source countries include Germany, Sweden, the Netherlands, Britain, Canada, the United States, Australia, and Japan. Popular destinations are widespread and growing, as indicated by the 2019 list of "the world's best places to retire", from *International Living* – an online magazine that for years has promoted global retirement. That list includes Panama, Costa Rica, Mexico, Ecuador, Malaysia, Colombia, Portugal, Peru, Thailand, and Spain (International Living 2019). Rating and ranking retirement destinations is highly subjective, as was pointed out by Forbes Magazine, which found very little concordance between the International Living list and one posted simultaneously by another group of international retirement experts, whose "best retirement city" selections were in the countries of Portugal, Ecuador, Slovenia, Vietnam, Brazil, Italy, France, and Indonesia (Eisenberg 2019).

Retirement migration in significant numbers first began in Europe, where as many as six million retirees are estimated to live outside their country of origin (Rojas et al. 2014). Popular destinations include southern France, Spain, and Italy (Warnes 2009; Benson 2010). An atypical migration pattern is the migration of retirees from the Netherlands to rural areas of Sweden. In this case the catalyst was a very active promotion by local governments in Sweden of rural life to people living in crowded urban settings (Eimermann 2015).

More recently, Asia has developed an increasingly prominent role in international retirement, because of its varied cultures, many warm locations, low cost of living and the availability of low-cost in-home assistance. Within Asia, Japan is by far the largest source of migrant retirees, with Chiang Mai, Thailand; Penang, Malaysia; Cebu, Philippines; and Bali, Indonesia being the most popular locations (Toyota and Xiang 2012). For Westerners retiring abroad, popular Asian destinations include Turkey, Thailand, India, Vietnam, and Malaysia (Warnes 2009;

International Living 2019; Eisenberg 2019; Fernández 2011). Thailand is unusual among international retirement destinations in that it attracts relatively more men than women, for many of the typical reasons cited above, but also due to an availability of younger Thai women as sexual partners (Warnes 2009).

Since the turn of the century, United States citizens have increasingly participated in the growing international retirement diaspora, with the estimated number of retired Americans living abroad growing by 17% between 2010 and 2015 and anticipated to mushroom further, as more baby boomers retire (Conlin 2009; Zamudio 2016). The predominant pattern of migration is from the U.S. and Canada to Mexico, Central America, and northern South America (Costanzo and von Koppenfels 2013). Mexico was the first main target and remains the most popular destination, with significant expatriate retiree populations developing first in the Lake Chapala region, San Miguel de Allende, and the Baja Peninsula (Truly 2002; Sunil et al. 2007).

## 1.2 Challenges in Quantifying International Retirement Migration

All experts agree that the number of persons moving from cooler to warmer countries in retirement is increasing; the challenge is turning that impression into an accurate count. Problems include persons who split time between residences in two countries, mixed nationality couples, persons who live abroad but maintain business addresses in the home country, younger persons who are working abroad (including an increasing number who can work remotely), and foreign born residents who came to the United States (or another developed country) for work and return "home" in later life, often to start businesses or to retire (Topmiller et al. 2011; Hayes 2015a). As a result of these and other complexities, governmental records from the U.S. or from Latin American countries are estimates and can vary widely (Schafran and Monkkonen 2011; Hayes 2015b; Dixon et al. 2006). One commentary on the demographics of migration to Mexico, for example, noted that in 1999–2000 the official Mexican and U.S. government estimates of US citizens living in Mexico varied by 300% (Topmiller et al. 2011).

Although concrete numbers are very difficult to grasp on the extent of retirement migration, it is clear that it is growing. In terms of U.S. citizens, the number of social security recipients receiving their monthly check abroad provides the best available estimate regarding the growth of the phenomenon, although for multiple reasons these figures underestimate the actual number of retirees living abroad (Warnes 2009). Between 1999 and 2017 the number of Americans having Social Security checks mailed to foreign addresses increased from 219,504 to 413,428 (Social Security Administration 2000, 2019). The majority of these were sent to Europe; however, as is illustrated by Fig. 1.1, the number sent to Latin America increased by 56% during those years. Noteworthy in Fig. 1.1 is growth after the economic downturn of 2008, illustrating the importance of financial issues in decisions to retire abroad.

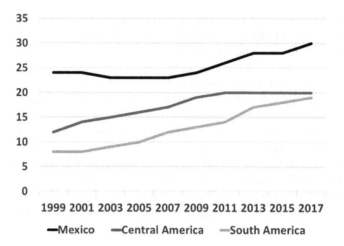

**Fig. 1.1** Number of U.S. citizens receiving Social Security checks (in thousands) in Latin America, 1999–2019, in thousands. These figures underestimate the number of retired Americans living abroad, as many have checks mailed to a U.S. address. The recent rise in South America is attributed largely to popularity of Ecuador and Colombia as retirement destinations. (Source: Social Security Administration, 1999–2018)

## 1.3 The Range of Motivations for International Migration in Retirement

According to Warnes, there are four general types of people who migrate internationally in retirement. They are: (1) persons who had moved from the "first world" to another country for work and return to their home country in retirement; (2) persons who moved from the "third world" to another country for work and return to their home country in retirement; (3) persons who move in retirement for amenity reasons from their home country to a location where they have not previously been permanent residents; and (4) family members who join individuals who are relocating or have relocated (Warnes 2009). According to this typology, most of the retirees discussed in this book are in groups 3 and 4, since they are citizens of the United States, Canada, and Western Europe who relocated to Latin America for retirement.

Motivating factors for this type of international retirement are quite varied. Among the principal motivators identified in surveys of migrant retirees in Latin America are: desire to live in a warmer or more temperate climate; search for a lower cost of living; desire for a better lifestyle than could be enjoyed at home (ranging from enjoying nature to having domestic help); familiarity with the destination country (by having previously visited or lived in the country); availability of a community of retirees from the home country; personal safety; economic and political stability of the destination country; disenchantment with some aspect of the home country (often either governmental or social trends); quality and cost of health care; and distance and ease of travel back to the home country (to visit family, take care of business, or obtain health care) (Rojas et al. 2014; Truly 2002; Topmiller et al. 2011; Hayes 2015b; Dixon et al. 2006; Davidson 2011; van Noorloos 2012).

Pickering et al., in a review of 44 studies of international retirement migration, identified four general motivational themes drawing retirees to move abroad: the destination (e.g., climate, natural and cultural amenities), the people (language, culture, available social networks), the cost of living, and the ease of movement (travel options, visa/residency requirements) (Pickering et al. 2018). Because so many of these reasons involve seeking greater amenities than are available in the migrant's country of origin, most retirement migration has been termed "lifestyle migration" (Hayes 2015b). While this definition fits many retirees, it omits the "push" factor of politics, living costs, or other concerns within the home country. Indeed, governmental policy and politics in source countries have at times served as a reason for emigration and a facilitator of that migration. A classic example is demonstrated in the movie *Casablanca*, where Rick's Café is portrayed as a gathering spot for Europeans who have emigrated to Morocco to escape World War II. Similarly, the election of a liberal U.S. President can motivate persons who are conservative to move abroad, and conversely the election of a conservative president can do the same for retirees who are liberal. In a study of retirees in the Lake Chapala region of Mexico, for example, Truly found that dissatisfaction with their home country for political, cultural, or economic reasons was a major reason why many had immigrated (Truly 2002).

## 1.4   Relationship Between Tourism and Retirement Migration

The relationship between tourism and retirement migration has been debated by researchers for decades. The concept of a tourism life cycle has been championed by Butler, who proposed that many tourist destinations progress from exploration, to development, to consolidation, and to stagnation, and that rejuvenation requires adaptation of the community (Butler 1980, 2011). Building on this framework, Foster and Murphy proposed that rebirth as a retirement destination was one of the paths by which communities could escape the decline associated with the stagnation phase, and that its development may occur contemporaneously with a community flourishing as a popular tourist site (Foster and Murphy 1991). Building on this work and our own background in gerontology, we have proposed a theoretical model that describes a community's evolution from tourism to retirement (Fig. 1.2). In our model, infrastructural development must proceed to make a community attractive to retirees, and that process starts with individuals purchasing second homes, proceeds to hosting increasing numbers of lifestyle/amenity retirees, and then, with aging of the retirement community, to the development of long-term care resources.

A challenge in both retiree migration and tourism is whether too much development can lead to diminution of the quality of life for local residents. One mechanism

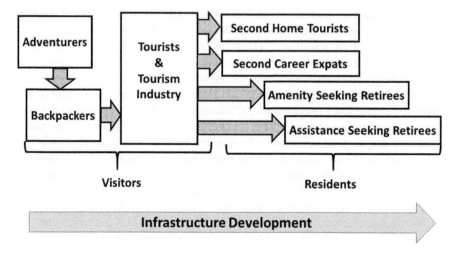

**Fig. 1.2** Theoretical model of the development of a community as a tourist site and subsequent progression to becoming a retirement destination. Infrastructure development is needed for a community to progress through each state of the model

by which this can develop is by overuse and degradation of environmental resources, such as water and open space; another can be through gentrification, which can force longstanding local residents to relocate to less costly neighborhoods in the outskirts of a community (Casado-Diaz 1999; Pacheco and Vallejo 2016).

## 1.5    Colonial Cities in Latin America as Retirement Destinations

Historic colonial cities are a popular destination for international retirees moving to Latin America (Schafran and Monkkonen 2011). Reasons for selecting these destinations include "charm, romance and splendor" (Pettiford 2013), affordability, and access to amenities such as restaurants, concerts, parks, and healthcare. Popular historic colonial cities for retirees include San Miguel de Allende, Mexico; Antigua, Guatemala; and Cuenca, Ecuador. Recently, the list has been expanding to include, among others, Cartagena, Colombia; Colonia, Uruguay; Casco Viejo, Panama; Granada, Nicaragua; and São Luis, Brazil (Pettiford 2013).

   A convenient way of identifying colonial cities that may have promise as retirement destinations is to look at those who have applied for and achieved UNESCO World Heritage status because of the quality and preservation of one or more historic neighborhoods. In Latin America there are 31 such cities: Sucre, Bolivia; Ouro Preto, Olinda, Salvador de Bahia, São Luís, and Diamantina, Brazil; Santa Cruz de

Mompox, and Cartagena, Colombia; Cienfuegos, Havana, Trinidad, and Camagüey, Cuba; Santo Domingo, Dominican Republic; Quito and Cuenca, Ecuador; Antigua, Guatemala; Mexico City, Oaxaca, Puebla, Guanajuato, Zacatecas, Morelia, Querétaro, Tlacotalpan, Campeche, and San Miguel de Allende, Mexico; Lima and Arequipa, Peru; Colonia del Sacramento, Uruguay; and Coro, Venezuela (United Nations Educational, Scientific, and Cultural Organization (UNESCO) n.d.).

In the academic literature, the first discussion of the potential of colonial cities as retirement destinations was published in 1994 in the Spanish language journal *Estudios Demográficos y Urbanos*. It was written by Leonard Plotnicov, an anthropologist at the University of Pittsburgh. His essay argued that medium-size cities in Mexico were at a disadvantage economically vis-à-vis commercial centers such as Mexico City and Monterrey and, as a result, were declining economically. He proposed that retirees from the United States were a large, growing potential resource that could be attracted by the charming architecture, relative peacefulness, low cost of living, climate, and recreational opportunities available in medium size Mexican colonial cities. While acknowledging that "*agringamiento*" (Americanization) could constitute a threat to the local culture, he recommended that Mexican cities consider implementing changes and policies that would make them attractive to retirees from the U.S., thereby providing them with an economic boost (Plotnicov 1994).

Of course, it takes more than charm and historic buildings to make a colonial city attractive to retirees. In a meeting with a demographer in Mexico, our conversation identified six criteria that differentiated San Miguel de Allende from other historic colonial cities that have been less successful in becoming retirement destinations:

- They must be clean. The houses should be freshly painted, and no garbage on the street.
- They must assiduously work to assure safety.
- They must invest money in activities (particularly in the arts) that would appeal to retirees.
- They must be away from crime-related routes such as petroleum pipelines in Mexico, or narcotic trafficking routes in much of Mexico and Central America.
- Their surrounding rural areas should be economically healthy. This has several advantages; most prominent are the availability of fresh produce for restaurants and a skilled population who are able to move to the city for work but not to escape poverty.
- They must promote themselves aggressively, focusing on both retirees and tourism. Tourists and retirees do have different needs, interests, and goals, but together they are complementary, and attracting tourists is often the first step to attracting retirees.

This list could constitute part of a blueprint for communities that want to follow Plotnicov's recommendation and become an international retirement destination to buoy their economies.

## 1.6    Cuenca, Ecuador and San Miguel de Allende, Mexico: Two Popular Destination Colonial Cities

Cuenca, Ecuador. Cuenca is a mid-sized city situated in the Andean Highlands of Ecuador. As of the 2010 census, there were 505,585 residents of the city (van Noorloos and Steel 2016). The weather has been described as "eternal spring" because of its consistently temperate climate, which is produced by being located near the equator and at 8400 feet of elevation in the Andean mountain range (Hayes 2015b). Another feature of the city is that four rivers run through the region, most notably the Tomebamba River. During the Inca period in the 15th and 16th centuries, the area was a major population center; the remaining Inca ruins and the historic nature of its colonial central city earned Cuenca UNESCO World Heritage status (Steel and Klaufus 2010).

A few years ago, the U.S. embassy estimated that around 4000 English speaking expatriates were living in Cuenca, partially due to publications like *International Living*, which touted Cuenca as a number-one retirement destination during several years between 2010 and 2016 (van Noorloos and Steel 2016). A study commissioned by the city of Cuenca itself found double that number, identifying growth from 2924 North Americans in 2001 to 8291 in 2010 (Álvarez et al. 2017). Since then, there has been continued growth of migrant retirees, who constitute up to 12% of the population 65 and older in Cuenca (Álvarez et al. 2017).

San Miguel de Allende, Mexico. San Miguel de Allende, a small city of around 130,000 inhabitants, is located within the state of Guanajuato in central Mexico. The city's climate is dry and temperate, owing to its location at 6200 feet of elevation in central Mexico. There is limited water in the area, as one of the small rivers of the municipality, el río Laja, grapples with pollution, and the nearby artificial lake is considered only useful for agricultural purposes (Cuellar Franco et al. n.d.).

Early in the eighteenth century, the town developed as a way station between the silver mines in northern Guanajuato state and Mexico City. By the 1740s, it was a flourishing city of 20,000 inhabitants, with an attractive central area in the Baroque and Neoclassic style that still stands today. In the subsequent two centuries it went through periods of growth and decline (López Morales 2008). Early in the twentieth century Jose Mojica, a Mexican film and opera star, and Manuel Toussaint, a Mexican art historian, began to champion San Miguel as an artistic and historic town (Covert 2017). Other artists and entrepreneurs followed, and by the early 1940s the city had become an international tourist destination (Covert 2017). The Mexican Government recognized San Miguel as a National Monument in 1982, and designated it a "Pueblo Magico" as part of a tourism initiative beginning in 2001. In 2005, UNESCO recognized the city as a World Heritage Site, based on the historic value of its city center (López Morales 2008).

Retirees soon followed the artists and tourists. Data collected between 2004 and 2005 found that the average American retiree in San Miguel had already been in the area for 7 years (Rojas et al. 2014). Recent estimates identify at least 10% of the population to be foreign-born, most of whom are retirees (Fernández 2011; Lizárraga-Morales 2010). With this population growth came concerns about impact – as far back as 2008 the New York Times reported that the city was on the precipice of losing its 'authenticity' due to the influx of retirees (Atkinson 2008).

## 1.7  Governmental Initiatives to Promote Retirement Immigration

In an attempt to attract international retirees and the income they can bring, local communities and national governments have implemented a variety of laws, regulations, and promotional campaigns. Among the earliest was the government of Malta, which soon after independence lowered the personal tax rate for British citizens, a move that over time led Malta to be a major retirement spot (Warnes 2009).

Often, governmental strategies specifically target areas that have seen little economic development or are declining, as has been done to promote internal migration in developed countries such as Australia and Canada (Carson et al. 2016). This was the strategy behind active promotion in the Netherlands rural settings in Sweden as "charming, bucolic" retirement destinations (Eimermann 2015). Unfortunately, such strategies often have unintended consequences. In the case of Dutch migrants to rural Sweden, over two-thirds returned home, often in disillusionment about the reality of living in a remote setting, and the move proved quite stressful for many couples, often because one liked the change and the other did not. Additionally, the retiree influx did not lead to as many employment opportunities for local residents as had been projected, in part because the retirees traveled frequently and in part because non-local persons who spoke Dutch were preferred for any work that involved interaction with the retirees (Eimermann 2016).

As with any business, international retirement is competitive; so, in response many countries have shortened and simplified visa requirements, developed discount programs for seniors, created tax incentives, paid for advertisements in the media, and created other incentives to lure transnational retirees (David et al. 2015). Typically, these initiatives involve partnerships between various business entities and national or local governments, with the goal of promoting economic development through "residential tourism." In several Southeast Asian countries, for example, retirement migration has been promoted in collaboration with the tourism industry, the real estate sector and medical care providers such as nursing homes and home services providers (Toyota and Xiang 2012). In the Philippines, retirement villages have been built and a government agency created to handle the influx of over 100,000 retirees, many of whom came from Japan because of the availability of affordable live-in domestic helpers and the ability to stretch pensions (Toyota and Xiang 2012). A similar program has been more recently developed in Malaysia, which has contributed to that country's ascendency as a preferred retirement destination (International Living 2019).

A similar process of collaboration between governments, real estate professionals, transnational businesses, the tourism industry, and lifestyle entrepreneurs is
ongoing in Latin America, termed "country marketization" (Pallares and Rollins-
Castillo 2019). Government plays a key role in this process through the passage of
incentives. The incentives put into place include liberalized visa requirements and a
variety of discounts. Concomitant with the incentives are financial solvency requirements – typically a stable monthly pension and a minimum level of savings.

Costa Rica was the earliest country in the Americas to create major incentives for
retiree migration, developing favorable policies in the in the 1980s (Dixon et al.
2006); however, because of the influx of migrants it received, these policies were
subsequently scaled back (Jackiewicz and Craine 2010). By early in the first decade
of this century it had been replaced as a leader in retirement-friendly regulation by
Panama, whose facilitative policies include excise tax exemptions on automobile
purchases or importation; a one-time import tax exemption on household goods of
up to $10,000; exemption from property taxes for 20 years on newly-built homes;
and discounts for seniors on movies, theater and public shows, restaurants, medical
care, energy and phone bills, hotels, and resorts (Dixon et al. 2006; Jackiewicz and
Craine 2010).

Mexico, too, has a number of policies that facilitate retirement migration. It
allows importation of household goods without import duties during first 6 months
of visa status; allows visitors and temporary residents to have foreign-plated cars;
has local property taxes that are far lower than in the U.S.; invests in enhanced security for foreigners; and through participation in the North American Free Trade
Agreement (NAFTA) has facilitated entry of companies such as Walmart and Costco
that provide products familiar to North American retirees (Dixon et al. 2006;
Pallares and Rollins-Castillo 2019).

Ecuador also has created policies to appeal to retirees. One of the biggest has
been unintentional – their conversion to the U.S. dollar as the national currency,
which has provided economic stability (Jackiewicz and Craine 2010). Temporary
residency for 2 years is relatively easy to obtain; permanent residency status can be
achieved after 3 years; and citizenship is available after three more years.
Additionally, the municipality of Cuenca has developed transportation, created a
city agency to assist tourists and retirees, increased public wireless internet availability, and put in place a security and assistance plan for foreigners, all of which
have helped it evolve as a desirable retirement destination (Pallares and Rollins-
Castillo 2019). In 2006 and 2007 the country launched a promotional campaign
aimed at putting it on the map as both a tourist and a retirement destination, which
has continued into this decade, even including the first ever television commercial
by a country during a Super Bowl (Thompson 2015).

Health care is an important issue for migrant retirees, and governmental policies
in both the home and destination countries impact a country's desirability. For
European migrants who relocate within the European Union, medical care is paid
for by reciprocity agreements between national health insurance plans among the
countries; however, satisfaction with local services is mixed and major health events

can still be disruptive. For migrants from the U.S., however, healthcare abroad is especially challenging. Medicare does not pay for health care outside the U.S. (Centers for Medicare and Medicaid Services 2019), and neither do most U.S.-based private insurance plans (Sloane et al. 2013). Proposals have been made for extending Medicare internationally (Warner and Jahnke 2001); but these have not gained traction and indeed seem further than ever from realistic consideration.

Both Mexico and Ecuador do provide some access on a paid basis to their national health services, as well as the availability of private health insurance plans, which typically are much less expensive that health insurance in the United States. Mexico allows persons who have temporary or permanent residency status to purchase enrollment in the *Instituto Mexicano de Seguro Social* (IMSS), the public healthcare system in that country, but satisfaction with services is often not good (Sloane et al. 2013). In Ecuador, all residents are required by law to have health insurance, and so even temporary residents are able to enroll in the public health system, the *Instituto Ecuatoriano de Seguridad Social* (IESS), which receives moderate to good reviews from immigrant retirees (Haines 2018).

## 1.8   Impact of Retirement Migration on Receiving Communities

Retirement migration and tourism are generally regarded as a clean, employment-generating form of economic development, in that "winners" within the city outnumber "losers" (van Noorloos 2012). Because this is viewed as an opportunity by many nations and communities, governments and corporate interests have often developed active promotional campaigns to attract retirees. The economic segments most commonly involved include the tourism industry, the real estate sector, and medical and long-term care providers (Toyota and Xiang 2012).

Limited research supports the notion that natives in receiving communities share the impression that retirement migration is good for receiving communities. In the Baja California community of Mulegé, for example, Topmiller, et al. conducted 25 formal interviews with Mexican citizens. Respondents were generally quite favorable in their impressions of the presence of foreigners, considering them a valuable resource, frequently noting philanthropic roles such as hosting charity events and fundraisers (Topmiller et al. 2011). Camacho, in a survey of local residents in Cuenca, found that 79% wanted to see more retirees come to live in the city (Camacho 2017).

On the other hand, some have argued that the economic development that often accompanies retiree migration largely benefits the elite within those communities, rarely trickling down to benefit the lower social classes (Rainer 2019). Indeed, it can be argued that even the promotional activities and governmental policies that have fostered both tourism and retiree migration originate from and benefit the upper classes in Latin American cities. Furthermore, the largely Caucasian retirees may fit into a color-conscious local hierarchy, thereby reinforcing long-established social and economic disparities (Hayes 2015a).

If retirees migrate in significant numbers, there is a potential for change within the local economy by driving up prices, particularly of real estate. Thus, landowners and real estate companies are among the big "winners" when a community becomes popular with migrant retirees (Rainer 2019). In the Lake Chapala region of Mexico, for example, where retirees have settled in large numbers over many decades, the economy has largely changed from farming to real estate. In the process, speculators made large profits by purchasing land from local residents at low prices, developing it, and selling it to retirees at much higher prices. Violation of local laws and traditions occurred in the process, such as sale of land that had traditionally been communally-owned (Bastos 2014).

When this process occurs within cities, an influx of migrants from elsewhere causes real estate prices to rise in desirable neighborhoods, which in turn leads to exodus of previous residents to less costly outlying neighborhoods, a process that is referred to as gentrification (Rainer 2019). The higher prices in the preferred areas tends to exclude certain low-income groups, including many younger persons, from the gentrified neighborhoods (van Noorloos and Steel 2016). One demographic study in Torrevieja, Spain, for example, reported that intense retiree migration had led to an economic boom but also "depopulation of the town center and the dispersal of the population" (Casado-Diaz 1999). When migration occurs in smaller numbers, it is still common for retirees to "appropriate" desirable spaces, such as living areas next to parks (Camacho 2017).

Cities with picturesque city centers, such as many UNESCO World Heritage Sites, are particularly prone to price escalation and gentrification, especially if the historic central area contains properties that are in need of renovation. This phenomenon that has been described in San Miguel de Allende, Mexico (Pacheco and Vallejo 2016); Cuenca, Ecuador (Hayes 2018a); Granada, Nicaragua (Reyes Rocha 2011); and Fez and Marrakesh, Morocco (Gil de Arriba 2011). Gentrification is common in tourism destinations and represents a kind of cultural appropriation, in which superficial artifacts of the old culture, such as architecture, music, and crafts may be preserved, but deeper levels, such as traditions, belief, and values, are eroded (George and Reid 2005). If sufficiently intense, it can result in the historical city center becoming so distinct from the remainder of the city as to practically be a foreign enclave (Reyes Rocha 2011).

While the economic effects of significant retiree migration are great, social structures appear less affected. In large measure this is because the retirees tend to socialize within their own enclaves rather than becoming integrated into local society (van Noorloos 2012; Rainer 2019). In Cuenca, Ecuador, this led to small foci of immigrant retirees, described as "hubs of gringolandia," which tended to be near parks, supermarkets, and restaurants, and the city center (van Noorloos and Steel 2016). Over time and in sufficient numbers, however, they can become a political force that actively promotes their own interests, often with more effectiveness than would be expected by their limited numbers (Covert 2017; Hayes 2018a).

One area of social impact of retirees that has been frequently praised is their involvement in voluntarism and philanthropy. Such activities are common in retirement communities, providing participants with both community and purpose (Haas 2013). As noted by Reyes Rocha, in a study in Granada, Nicaragua, "foreigners start projects based on their appreciation and understanding (of local problems)" (Reyes Rocha 2011). However, whether and to what extent these activities truly improve the well-being of the local population is much debated, and likely varies by location, by activity, and by the extent to which they are done in partnership with members of the local community.

A related issue is whether lifestyle migration to less developed countries involves a degree of exploitation, which has been referred to as "postcolonialism" (Rainer 2019; Benson 2013). One example is North American retirees to rural Boquete, Panama, whose presence initiated changes in the economy and the social structure of the area, including the retirees hiring many local residents as housekeepers, gardeners, and other domestic workers, thus underscoring the privileged status of the newcomers. The process did bring about economic development, but vast disparities between the retirees and the local residents persisted. In this context such activities as volunteering, charitable work, and philanthropy by the economically better-off retirees can be viewed as not addressing the underlying social disparities but rather as an expression of privilege and power, particularly when the charitable activities primarily benefit the immigrant retirees (Benson 2013).

Significant retiree migration transforms local communities into "transnational spaces," with the migrants creating social and cultural ripple effects that can be wide-reaching, disrupting traditional economic and social patterns (van Noorloos 2012). This phenomenon has been referred to as "dispossession," and is viewed by some as having a coercive component. Bastos, for example, describes the disruption of indigenous communities that had been in the Lake Chapala area "since time immemorial" by developmental pressure associated with the area becoming an international retirement destination. Usually the process is peaceful, and development is fostered by "neoliberal" governmental policies and actions. However, occasionally, the development process has led to open conflict between supporters and opponents (Bastos 2014). The results include an erosion of social cohesion within the local community and reinforcement of differences between the elite, who have global mobility, and many local residents, who travel only to escape poverty or danger (van Noorloos 2012).

Another theoretical perspective on this issue involves whether and to what extent retiree immigration impinges on the local population's "right to the city" – a concept that has been advanced in Latin America by urban social movements and policy forums (Fernández 2011). From this viewpoint, conversion of historic or culturally valuable areas of a city to tourism or retiree immigration can be seen as a type of cultural appropriation of social spaces, driven in part by an uneven power imbalance between the wealthier immigrants and the resident local inhabitants. Entry of businesses owned by non-natives of the city, ranging from single-site restaurants to international chains such as Starbucks, often accompany large scale retirement migration, further disrupting the traditional social spaces, as well as the local economy (Fernández 2011).

The immigrant retirees themselves are often aware of the role of their activities or of power and income inequalities in changing their host cities, though some may not appreciate the extent to which their taking advantage of global wealth imbalances causes displacement of local residents or appropriation of local spaces. Those who are particularly disturbed by the belief that they are participating in "white privilege" seek moral responses through being "good neighbors," by paying higher prices than local residents and being respectful of local traditions (Hayes 2018b).

In sum, consensus on the value and impact of retirement migration on receiving communities is quite mixed. Among researchers, no consensus exists regarding whether and to what extent the positives outweigh the negatives, or vice versa (Rojas et al. 2014). One potential reason for this is that the majority of studies have been done by academics from developed countries and have focused on the retirees themselves; so, relatively little is known about the opinions of residents of the receiving communities. Additionally, research is quite limited in regard to both the economic and particularly the social/environmental impact of retirement migration – despite the recommendation that "the social effects are so significant that they should be studied before anything else" (Brunt and Courtney 1999).

## 1.9 Gaps in the Research on International Retirement Migration in Latin America

In spite of growing interest in the topic, a number of gaps exist in the literature on international retirement migration to Latin America. As noted above, the immigrant retirees have been subjected to more research than the local populations (Schafran and Monkkonen 2011). Additionally, there is a dearth of comparative studies "that compare data across different spaces or types of built environments" (Schafran and Monkkonen 2011).

Much exploration is also needed on the relationship between international tourism and international retirement. Research is needed to explore to what extent and in what ways international retirees differ from tourists in terms of needs, resource use, and behavior (Butler 2011). It is well acknowledged that countries such as Mexico, Costa Rica, Panama, and Ecuador have actively promoted tourism and, as a secondary consequence, retiree migration; however, the connection between these efforts and increases in permanent residents is relatively unexplored (Schafran and Monkkonen 2011).

Much more needs to be studied about the social impact of retirement migration. According to Schafran and Monkkonen, "in-depth research is needed on the nature of the impacts of U.S. migrants on the places … to which they are moving," including "the way in which migrants are involved in the place to which they move"

(Schafran and Monkkonen 2011). More work is needed to determine to what extent migration is a form of exploitation or, as some have suggested, of neo-colonialism, in which the migrants by virtue of their community of origin and superior economic resources are in a privileged position and, therefore, may create undesired changes in the destination community (Fernández 2011).

Another gap is research related to health care, and particularly long-term care and end-of-life care. As noted by Schafran and Monkkonen, the high cost of nursing home care and similar services in the U.S. makes Latin America "particularly attractive" (Schafran and Monkkonen 2011); however, little research has been done in this incipient area.

Another area needing more study is the variation in migrant retirees (Schafran and Monkkonen 2011). Migrant retirees are a diverse group. Furthermore, it is well acknowledged that populations become more diverse with age (Sloane et al. 2014). so migrant retirees would be expected to be a highly varied group not just in terms of motivation or goals, but also in terms of health characteristics and needs. However, subgroups and their characteristics, needs, activities, and impacts have not been adequately explored (Schafran and Monkkonen 2011). For example, virtually nothing has been written about the problems, activities, and impact of immigrant retirees who are disadvantaged through physical or mental illness (Warnes 2009).

Finally, little is known about the long-range trends in terms of retiree migration and their medium to long-term implications on the host societies. Will retirement immigration be a tsunami or remain the relative trickle that it is now? How many migrants, and which ones, will remain in their country of relocation, requiring long-term care services? And how many, and which ones, will eventually move on or return to the home country?

Thus, this field is ripe for further research. And if retirement migration continues to grow, that research will be essential to help prepare host communities to take steps to maximize the positive impacts and minimize the adverse consequences of this relatively new population movement.

## 1.10  Goals of the Research Described in This Book

As noted in the previous section, the literature on retirement migration is relatively modest and tends to focus on the retirees themselves. Particularly sparse are research data on the impact of retiree migration on these cities, on health care issues, and comparative studies of more than one retirement location. The study presented in this book sought to help address these gaps by gathering data on both the retirees and the local population of Cuenca, Ecuador and San Miguel de Allende, Mexico, with a focus on the impact of the retirees from the point of view of the local residents. The specific goals and methods of the research are described in the introduction and the appendix; the remaining chapters of this book present results.

# References

Álvarez, M. G., Guerrero, P. O., & Herrera, L. P. (2017). Estudio sobre los impactos socio-económicos en Cuenca de la migración residencial de norteamericanos y europeos: Aportes para una convivencia armónica local. In *Informe Final*. Cuenca: Avance Consultora.

Atkinson, J. (2008, November 20). The New Old Mexico: In San Miguel de Allende, Mexico, Worry That New Americans Strain the Town's Charm. *New York Times*. https://www.nytimes.com/2008/11/21/greathomesanddestinations/21expat.html. Accessed 24 June 2019.

Bastos, S. (2014). Territorial dispossession and indigenous rearticulation in the Chapala Lakeshore. In M. Janoschka & H. Haas (Eds.), *Contested spatialities, lifestyle migration and residential tourism* (pp. 63–75). Abingdon: Routledge.

Benson, M. C. (2010). The context and trajectory of lifestyle migration: The case of British residents of Southwest France. *Journal of European Societies, 12*(1), 2010. https://doi.org/10.1080/14616690802592605.

Benson, M. C. (2013). Postcoloniality and privilege in new lifestyle flows: The case of North Americans in Panama. *Mobilities, 8*(3), 313–330.

Brunt, P., & Courtney, P. (1999). Host perceptions of sociocultural impacts. *Annals of Tourism Research, 6*(3), 493–514.

Butler, R. W. (1980). The concept of the tourist area cycle of evolution: Implications for management of resources. *Canadian Geographer, 24*, 5–12.

Butler, R. W. (2011). Tourism area life cycle. In *Contemporary tourism reviews*. Oxford: Goodfellow Publishers Limited.

Camacho, M. R. Z. (2017). Migración Norte – Sur: Inmigrantes jubilados estadounidenses y su proceso de inserción en Cuenca. *Master's Thesis: University of Cuenca*. Cuenca: Universidad de Cuenca. http://dspace.ucuenca.edu.ec/handle/123456789/28787. Accessed 24 June 2019.

Carson, D. A., Cleary, J., de la Barre, S., Eimermann, M., & Marjavaara, R. (2016). New mobilities – New economies? Temporary populations and local innovation capacity n sparsely populated areas. In A. Taylor, D. B. Carson, P. C. Ensign, L. Huskey, & R. O. Rasmussen (Eds.), *Settlements on the edge: Remote human settlements in developed nations* (pp. 178–200). Cheltenhan/Northampton: Edward Elgar Publishing.

Casado-Diaz, M. A. (1999). Socio-demographic impacts of residential tourism: A case study of Torrevieja, Spain. *International Journal of Tourism Research, 1*(4), 223–237.

Centers for Medicare and Medicaid Services. (2019). *Medicare and you – 2019*. https://www.medicare.gov/Pubs/pdf/10050.pdf. Accessed 31 May 2019.

Conlin, M. (2009, July 2). Retirement: Why Panama is the new Florida. *Business Week*. http://www.businessweek.com/magazine/content/09_28/b4139054352321.htm. Accessed 28 June 2017.

Costanzo, J., & von Koppenfels, A. (2013). Counting the uncountable: Overseas Americans. *Migration Policy Institute*. http://www.migrationpolicy.org/article/counting-uncountable-over-seas-americans. Accessed 28 June 2017.

Covert, L. P. (2017). *San Miguel de Allende: Mexicans, Foreigners, and the making of a World Heritage Site*. Lincoln/London: University of Nebraska Press.

Cuellar Franco, J. L., Hidalgo, M. G. A., & Sánchez, J. C. M. (n.d.). Allende. *Enciclopedia de los Municipios y Delegaciones de Mexico, Ayuntamiento de Allende*. http://siglo.inafed.gob.mx/enciclopedia/EMM11guanajuato/municipios/11003a.html. Accessed 10 July 2019.

David, I., Eimermann, M., & Akerlund, U. (2015). An exploration of lifestyle mobility industry. In K. Torkington, I. David, & J. Sardinha (Eds.), *Practicing the good life: Lifestyle migration in practices* (pp. 138–116). Newcastle upon Tyne: Cambridge Scholars Publishing.

Davidson, L. (2011, April 7). A great retiree migration abroad is not so far fetched. *Forbes Personal Finance*. http://www.forbes.com/sites/financialfinesse/2011/04/07/a-great-retiree-migration-abroad-is-not-so-far-fetched-2/. Accessed 31 May 2019.

Dixon, D., Murray, J., & Gelatt, J. (2006). *America's emigrants: U.S. retirement migration to Mexico and Panama*. Washington, DC: Migration Policy Institute.

Eimermann, M. (2015). Promoting Swedish countryside in the Netherlands: International rural place marketing to attract new residents. *European Urban and Regional Studies, 22*(4), 398–415.

Eimermann, M. (2016). Two sides of the same coin: Dutch rural tourism entrepreneurs and countryside capital in Sweden. *Journal of Rural Society, 25*(1), 55–73.

Eisenberg, R. (2019, January 4). The top 10 places in the world to retire: 2 new lists. *Forbes.* https://www.forbes.com/sites/nextavenue/2019/01/04/the-top-10-places-in-the-world-to-retire-2-new-lists/#5c8a93c156bc. Accessed 20 June 2019.

Fernández, I. G. (2011). The Right to the City as a conceptual framework to study the impact of North-South Migration. *Recreation and Society in Africa, Asia and Latin America, 2*(1), 3–33.

Foster, D. M., & Murphy, P. (1991). Resort cycle revisited: The retirement connection. *Annals of Tourism Research, 18*, 553–567.

George, E. W., & Reid, D. G. (2005). The power of tourism: A metamorphosis of community culture. *Journal of Tourism and Cultural Change, 3*(2), 88–107.

Gil de Arriba, C. G. (2011). El turista de las mil y una noches. Turismo residencial en Marruecos: transformación funcional y simbólica del patrimonio arquitectónico y del territorio. In T. Mazón, R. Huete, & A. Mantecón (Eds.), *Construir una nueva vida: Los espacios del turismo y la migración residencial* (pp. 203–223). Santander: Editorial Milrazones.

Haas, H. (2013). Volunteering in retirement migration: Meanings and functions of charitable activities for older British residents in Spain. *Ageing & Society, 33*, 1374–1400.

Haines, B. (2018). Do I need health insurance in Ecuador? What residents & travelers need to know. *Gringos Abroad.* https://gringosabroad.com/health-insurance-in-ecuador/. Accessed 22 June 2019.

Hayes, M. (2015a). 'It is hard being the different one all the time': Gringos and racialized identity in lifestyle migration to Ecuador. *Ethnic and Racial Studies, 38*(6), 943–958.

Hayes, M. (2015b). Moving south: The economic motives and structural context of North America's emigrants in Cuenca, Ecuador. *Mobilities, 10*(2), 267–284.

Hayes, M. (2018a). *Gringolandia: Lifestyle migration under late capitalism.* Minneapolis/London: University of Minnesota Press.

Hayes, M. (2018b). The gringos of Cuenca: How retirement migrants perceive their impact on lower income communities. *Royal Geographical Society: Area Journal, 50*(4), 1–9.

International Living. (2019). The world's best places to retire in 2019. *International Living.* https://internationalliving.com/the-best-places-to-retire/. Accessed 20 June 2019.

Jackiewicz, E. L., & Craine, J. (2010). Destination Panama: An examination of the migration-tourism-foreign investment nexus. *RASALA, 1*(1), 5–29.

Lizárraga-Morales, O. (2010). The US citizens' retirement migration to Los Cabos, Mexico. Profile and social effects. *Recreation and Society in Africa, Asia & Latin America, 1*(1), 75–92.

López Morales, F. J. (2008). *Protective Town of San Miguel and the Sanctuary of Jesus de Nazareno de Atotonilco.* UNESCO, World Heritage Center. http://whc.unesco.org/en/list/1274/. Accessed 24 June 2019.

Pacheco, M. I. F., & Vallejo, M. P. G. (2016). Entre lo local y lo foráneo: Gentrificacion y discriminación en San Miguel de Allende, Guanajuato. *Revista Legislativa de Estudios Sociales y de Opinion Publica, 9*(18), 183–206.

Pallares, A., & Rollins-Castillo, L. J. (2019). Lifestyle migration and the marketization of countries in Latin America. In A. Feldmann, X. Bada, & S. Schütze (Eds.), *New migration patterns in the Americas* (pp. 171–199). Cham: Palgrave Macmillan.

Pettiford, K. (2013, January 22). Five top colonial cities in the Americas. *US News and World Report Money.* http://money.usnews.com/money/blogs/on-retirement/2013/01/22/5-top-colonial-cities-in-the-americas. Accessed 30 May 2019.

Pickering, J., Crooks, V. A., Snyder, J., & Morgan, J. (2018). What is known about the factors motivating short-term international retirement migration? A scoping review. *Population Ageing*, 1–17.

Plotnicov, L. (1994). El atractivo de las ciudades medias. *Estudios Demográficos y Urbanos., 9*(2), 283–301.

Rainer, G. (2019). Amenity/lifestyle migration to the Global South: Driving forces and socio-spatial implications in Latin America. *Third World Quarterly, 40*(7), 1359–1377.

Reyes Rocha, F. A. (2011). *Exploring multiple Modernities, case study for Granada, Nicaragua: The emergence of a localized modernity in a context of international retirement migration.* Thesis, Wageningen University. Wageningen: Wageningen University.

Rojas, V., LeBlanc, H. P., & Sunil, T. S. (2014). US retirement migration to Mexico: Understanding issues of adaptation, networking, and social integration. *Journal of International Migration and Integration, 15*(2), 257–273.

Schafran, A., & Monkkonen, P. (2011). Beyond Chapala and Cancún: Grappling with the impact of American migration to Mexico. *Migraciones Internacionales, 6*(2), 223–258.

Sloane, P. D., Cohen, L. W., Haac, B. E., & Zimmerman, S. (2013). Health care experiences of U.S. retirees living in Mexico and Panama: A qualitative study. *BMC Health Services Research, 13*, 411.

Sloane, P. D., Warshaw, G. A., Potter, J. F., Flaherty, E., & Ham, R. J. (2014). Principles of primary care of older adults. In *Primary care geriatrics* (5th ed.). Chicago: Mosby-Yearbook.

Social Security Administration. (2000). *Annual statistical supplement to the social security bulletin, 1999.* Washington, DC: United States Congress, Office of Retirement and Disability Policy, Office of Research, Evaluation, and Statistics.

Social Security Administration. (2019). *Annual statistical supplement to the social security bulletin, Dec 2017.* Washington, DC: United States Congress, Office of Retirement and Disability Policy, Office of Research, Evaluation, and Statistics.

Steel, G., & Klaufus, C. (2010). *Displacement by/for development in two Andean cities.* Paper presented at the 2010 Congress of Latin American Studies Association, Toronto, Canada. https://www.researchgate.net/publication/315542207_Displacement_byfor_development_in_two_Andean_cities. Accessed 26 July 2019.

Sunil, T. S., Rojas, V., & Bradley, D. E. (2007). United States' international retirement migration: The reasons for retiring to the environs of Lake Chapala, Mexico. *Ageing & Society, 27*(4), 489–510.

Thompson, C. (2015, January 29). Ecuador to become first foreign government to advertise during Super Bowl. *CNN.* https://www.cnn.com/travel/article/ecuador-super-bowl/index.html. Accessed 22 June 2019.

Topmiller, M., Conway, F. J., & Gerber, J. (2011). US migration to Mexico: Numbers, issues, and scenarios. *Mexican Studies/Estudios Mexicanos, 27*(1), 45–71.

Toyota, M., & Xiang, B. (2012). The emerging transnational "retirement industry" in Southeast Asia. *International Journal of Sociology and Social Policy, 32*(11/12), 708–719.

Truly, D. (2002). International retirement migration and tourism along the Lake Chapala Riviera: Developing a matrix of retirement migration behavior. *Tourism Geographies: International Journalism of Tourism Space, Place, and Environment, 4*(3), 261–280.

United Nations Educational, Scientific, and Cultural Organization (UNESCO). (n.d.). *World heritage list.* https://whc.unesco.org/en/list/. Accessed 14 June 2019.

van Noorloos F. (2012). Whose place in the sun? Residential tourism and its implications for equitable and sustainable development in Guanacaste, Costa Rica. Delft: Uitgeverij Eburon.

van Noorloos, F. K., & Steel, G. (2016). Lifestyle migration and socio-spatial segregation in the urban(izing) landscapes of Cuenca (Ecuador) and Guanacaste (Costa Rica). *Habitat International, 54*, 50–57.

Warner, D. C., & Jahnke, L. R. (2001). Toward better access to health insurance coverage for US retirees in Mexico. *Salud Pública de México, 43*(1), 59–66.

Warnes, T. A. (2009). International retirement migration. In P. Uhlenberg (Ed.), *International handbook of population aging.* New York/Berlin/Heidelberg/Dordrecht: Springer.

Zamudio, M. (2016, December 29). Growing number of Americans are retiring outside the US. *USA Today.* https://www.usatoday.com/story/money/personalfinance/2016/12/29/growing-number-americans-retiring-outside-us/95885494/. Accessed 31 May 2019.

# Chapter 2
# Living as a Retired Immigrant in Cuenca, Ecuador and San Miguel de Allende, Mexico

**Johanna Silbersack, Philip D. Sloane, and Karla Jimenez-Magdaleno**

With minimal internet sleuthing, it is possible to uncover extensive communities composed of persons of retirement age comparing notes, advice, and news from countries and cities within the global South, including destinations in Spain, Thailand, South America, Mexico, Malta, and others. Veritable fountains of local information, primarily from the retired expatriates' perspectives, can be found by visiting sites like *Gringo Post*, an online forum for retirees in Ecuador, or *Atención San Miguel*, a bilingual, weekly newspaper set in San Miguel de Allende, Mexico.

These sites, and many others, function as stomping grounds for current and future retirees who make up the North-South international retirement migration phenomenon. *Gringos Abroad* (Ecuador), *San Miguel Civil List* (yahoo & facebook groups), *Expat Exchange*, and many others serve as a means to connect and exchange information on movie screenings, the availability of diabetes support groups, a request for a sewing machine, or access to local public pools within specific cities. The forums offer a snapshot into some of the daily going-ons and needs of the retired immigrant community. For example, Fig. 2.1 is a photo of a newspaper clipping from *Atención San Miguel*; it describes a selection of lectures, concerts, and art events that can be found in San Miguel de Allende, many of which (although not all) are held in English as events for the foreign community.

In order to understand the impact that international retirement migration may have on communities, it is important to have a sense of the expatriates who have chosen to retire in this way. Although international retirement is still far from mainstream, the phenomenon has been growing steadily since the advent of the internet and the resulting increased globalization in the early 2000s, to the point that as many as 5 million retirees may be currently living abroad (van Noorloos 2012).

J. Silbersack · P. D. Sloane (✉) · K. Jimenez-Magdaleno
University of North Carolina at Chapel Hill, Chapel Hill, NC, USA
e-mail: philip_sloane@med.unc.edu

© Springer Nature Switzerland AG 2020        19
P. D. Sloane et al. (eds.), *Retirement Migration from the U.S. to Latin American Colonial Cities*, International Perspectives on Aging 27,
https://doi.org/10.1007/978-3-030-33543-4_2

**Fig. 2.1** A photo of the weekly newspaper, Atención San Miguel, published on March 16th, 2018. Atención is a bilingual (Spanish & English) newspaper begun by La Biblioteca de San Miguel de Allende, a local library nonprofit. Both endeavors have longstanding relationships with expatriates

Articles from mainstream media, including the *Washington Post, Forbes,* the *New York Times,* and others (Sheridan 2019; Talty 2017; Opdyke 2008; Emling 2010), offer often glowingly favorable reflections on the trend and advice for those considering a move. These media, and others like *International Living,* generally present living internationally as a healthy, affordable alternative to aging in place – with articles titles such as "Enjoying an Active Life in Costa Rica" and rankings of top destinations for affordable health care, desirable climates, and other appealing features (International Living 2019a; Ocampo 2019).

These factors, combined with the opportunity for community building online, have contributed to the rise of international retirement from the United States and Canada. U.S. social security data show an increase in checks sent to retired workers abroad from approximately 220,000 in 1999 to 413,000 in 2017 (Social Security Administration 2000, 2019). Considering that many retirees have U.S. addresses, these numbers are likely underreporting the true extent of the trend (Schafran and Monkkonen 2011).

Previous literature on retirement migration within Latin America has spent a substantial amount of time investigating the retirees themselves – including their demographics, desires, motivations, lifestyle upon moving, community integration, and decision-making process (Kiy and McEnany 2010a; Dixon et al. 2006; Rojas et al. 2014; Truly 2002; Otero 1997; Sunil et al. 2007; Lizárraga-Morales 2009; Li Ng and Ordaz Díaz 2012; van Noorloos and Steel 2016). The motivations for retirees range far and wide, from pursuing an adventure, a new culture and lifestyle, to escaping conditions within home countries – whether financial, political, or other.

In this chapter, we further the study of international retirement by contextualizing the characteristics of retirees within the two cities of Cuenca, Ecuador and San Miguel de Allende, Mexico. We will also, in this and other chapters of this book, expand on prior research by explicitly exploring the perceptions of retired immigrants regarding their own impact on the cities.

We drew data from two online surveys (one in each city) plus observations and interviews conducted by the research team in the two study cities. An online survey was chosen because of its efficiency, its ability to reach the majority of retirees through internet dissemination channels, and our ability to make modifications in real time if needed. The survey was restricted to persons aged 55 and older who moved to either San Miguel or Cuenca from the U.S., Canada, or Europe after the age of 50. The survey data are supplemented from results of the research team's networking within the retiree communities, both formally and informally, during data collection visits to Cuenca in June and San Miguel de Allende in July of 2018.

The research team used a number of methods to disseminate the survey, including networking with retired expatriates in each city, posting flyers in spaces well-known to have a substantial expatriate presence, and promoting the survey on popular online forums and listservs including *Gringo Post* and *Cuenca High Life* in Cuenca, and *Atención San Miguel*, *SMA_List* and *SMA_CoolList* in San Miguel de Allende. These methods of dissemination were chosen due to retired expatriates' ubiquitous use of the internet (Kiy and McEnany 2010a; Bantman-Masum 2015).

The full survey and a detailed accounting of the methods can be found in the Appendix.

## 2.1  Characteristics of Expat Retirees

We obtained 749 responses from retired immigrants, 424 from Cuenca, Ecuador and 325 from San Miguel de Allende. The majority of respondents were from North America, with 88.9% identifying the United States and 8.4% identifying Canada as their country of origin. This mirrors other accounts of the immigrant retiree populations within these two cities. A study commissioned by Cuenca's municipality found that up to 93% of their retired immigrants came from the U.S. and Canada (Álvarez et al. 2017). Comprehensive comparison data are not available for San Miguel de Allende; however, by 2011 it was projected that 10% of the municipality was composed of retired expatriates, 90% of whom came from the U.S. and Canada (Fernández 2011).

**Table 2.1** Comparison between demographic characteristics of survey respondents in Cuenca, Ecuador (n = 424) and San Miguel de Allende, Mexico (n = 325)[a]

| Variable | | Mean (SD) or Percent (n) | | | P-Value for difference between cities |
|---|---|---|---|---|---|
| | | Overall | Cuenca | San Miguel de Allende | |
| **Age** | Average | 68.97 (SD = 6.4) | 67.87 (SD = 5.9) | 70.41 (SD = 6.79) | <0.001[b] |
| | 55–59 | 7.0% (51) | 8.3% (34) | 5.4% (17) | <0.001[c] |
| | 60–64 | 18.1% (132) | 19.2% (79) | 16.8% (53) | |
| | 65–69 | 27.9% (203) | 34.2% (141) | 19.6% (62) | |
| | 70–74 | 28.7% (209) | 27.4% (113) | 30.4% (96) | |
| | 75+ | 18.3% (133) | 10.9% (45) | 27.9% (88) | |
| **Gender** | Male | 49.0% (361) | 54.6% (228) | 41.7% (133) | <0.001[c] |
| | Female | 50.3% (371) | 44.3% (185) | 58.3% (186) | |
| | Prefer not to answer | 0.7% (5) | 1.2% (5) | 0% (0) | |
| **Education** | High school or less | 5.7% (37) | 7.6% (28) | 3.3% (9) | 0.122[c] |
| | 2-year college | 13.8% (89) | 13.6% (50) | 14.2% (39) | |
| | 4-year college | 27.0% (174) | 27.4% (101) | 26.6% (73) | |
| | Post college | 53.4% (344) | 51.5% (190) | 56.0% (154) | |
| **Income** | >$1000 | 5.9% (36) | 7.1% (25) | 4.2% (11) | <0.001[c] |
| | $1000–$3000 | 48.4% (297) | 55.9% (198) | 38.1% (99) | |
| | <$3000 | 45.8% (281) | 37.0% (131) | 57.7% (150) | |

[a]Numbers vary due to non-response
[b]Statistic computed using two independent sample t-test
[c]Statistic computed using Chi Square or, for ordinal responses, Cochran-Mantel-Haenszel Chi-Square

Table 2.1 summarizes demographic characteristics of the respondents. Respondents from San Miguel de Allende and Cuenca were statistically different in almost all domains, with the exception of educational levels. Retirees across both cities reported high levels of education, with the majority in each city having advanced degrees, and less than 20% of respondents reported less than a Bachelors' degree. Respondents in San Miguel de Allende tended to be older, were more likely to identify as female, and on average had higher incomes than respondents in Cuenca. Although the mean age of all respondents was 69 years (SD 6.4), they ranged from 55 (which was the minimum age to qualify for our survey), to 92 years old. A large minority of respondents (27.9%) were aged 75 and older in San Miguel de Allende, whereas far fewer of Cuenca's respondents (10.9%) were in this older age group (p < 0.001). This demonstrates San Miguel's more established history as an expatriate destination and indicates that the numbers of retirees choosing to age in place there are substantial.

As mentioned above, retirees in San Miguel de Allende reported higher incomes than those in Cuenca, with 58% of the expatriates in SMA living on incomes of

$3000 or more a month, compared to Cuenca's majority of retirees living on incomes between $1000 and $3000 a month. This reinforces the narratives uncovered through interviews and fieldwork within each site that reflected the fact that retirees in Cuenca often used the term 'economic refugee' to describe themselves (Hayes 2018a). This seems to be true not only of our own sample, but within other research investigations as well. A study of 375 retirees in San Miguel in the mid-2010s reported average annual salaries of $56,000–$60,999 (Rojas et al. 2014), versus Cuenca's retirees average annual income of approximately $28,000 for a single household and $38,700 for households of two or more persons (Álvarez et al. 2017).

The majority of the retirees who responded to our survey were full time residents of the two cities; 74% had obtained permanent residency at the time of our survey, and an additional 2% had invested the time and effort to become citizens. The majority of respondents who did not have a permanent residency had temporary residency status (15%), with only a handful (6%) on a tourist visa. On average, retirees spent 10 months of the year in Mexico or Ecuador, and presumably 2 months traveling abroad or returning to their home country.

### 2.1.1  Reasons for Retiring Abroad

Before considering a permanent move, retirees must first learn about the destination community, either virtually or in person. Common methods include travel, internet websites and blogs, print materials, and conversations with family, friends, or acquaintances already living in the area. San Miguel de Allende and Cuenca differed significantly in how our survey respondents reported deciding on each city. In San Miguel de Allende, which has been a retirement destination for far longer than Cuenca, retirees rely far more heavily upon friends already living in the area, as well as prior travel to the location. Alternatively, retirees to Cuenca relied more heavily on websites and blog resources online ($p < 0.001$). This use of internet resources by retirees in Cuenca has been confirmed by other research investigating decision-making factors for expatriates, which found that 71% of retirees living there used internet research and digital magazines in order to inform their decision (Pacheco 2016).

### 2.1.2  Motivating Factors for International Relocation

The motivating factors for retirees to make the decision to retire internationally have been described by multiple researchers, and a number of migration theories attempt to explain the motivations behind such moves. One such theory that has been commonly referenced and used is the idea of "push" and "pull" factors on the individual level, which encourage people to move based on both incentives abroad and deterrents to stay within their home country (van Noorloos 2012; Truly 2002;

**Table 2.2** Factors Associated with decision to relocate internationally in retirement to Cuenca, Ecuador (n = 424) and San Miguel de Allende, Mexico (n = 325)[a]

| Factor | Mean (SD) | | | P-Value for difference between cities |
| --- | --- | --- | --- | --- |
| | Overall | Cuenca | San Miguel de Allende | |
| **Decision making process** | | | | |
| Friends already living there | 12.6% (91) | 8.3% (34) | 18.3% (57) | <0.001[b] |
| Books/magazines | 7.5% (54) | 8.1% (33) | 6.7% (21) | |
| Websites/blogs | 22.6% (163) | 35.6% (146) | 5.5% (17) | |
| Previous travel | 45.3% (327) | 37.8% (155) | 55.1% (172) | |
| Other | 12.1% (87) | 10.2% (42) | 14.4% (45) | |
| **Primary motive for relocating** | | | | |
| Affordable retirement or health care | 42.5% (307) | 52.3% (214) | 29.7% (93) | <0.001[b] |
| Amenities (lifestyle, climate, culture, adventure) | 42.4% (306) | 33.3% (136) | 54.3% (170) | |
| Get away from home country | 8.5% (61) | 8.1% (33) | 9.0% (28) | |
| Other | 6.7% (48) | 6.4% (26) | 7.0% (22) | |

[a]Numbers vary due to non-response
[b]Statistic computed using Chi-Square procedure

Álvarez et al. 2017; López 2017). For example, pull factors would be what the migrant positively attributes to their destination country; in the case of retirees, this often can include such elements as cost of living, availability of real estate, climate, and culture. Push factors are those that promote leaving a home setting; these include such factors as poor quality of life, problems with national politics, a domestic breakup, or lack of freedom. For retirees, 'push' factors often concern the cost of living in their country of origin or disgruntlement with political or social issues. This framework is useful on a micro scale, to understand what motivates one individual or couple to make the decision; however, the model has faced criticism for its focus on the individual and the exclusion of societal and cultural factors that can influence migration trends (López 2017).

Table 2.2 displays the results of our online surveys in Cuenca and San Miguel de Allende regarding reasons why immigrant retirees have been attracted to these cities. The subsequent sections describe the key factors, drawing from the literature as well as from our survey data and field interviews.

**Post-recession Cost of Living** The market crash and recession of 2008 were often related to as a major reason that retirement within expatriates' home countries became untenable for many immigrant retirees living on modest fixed incomes – a finding that has been previously described in research on Cuenca (Pacheco 2016). Survey respondents and retired expats we interviewed spoke of losses of employment, savings, or homes causing their long-term retirement plans to change drastically. For these reasons, expatriates in Cuenca particularly described themselves as 'economic refugees', a term increasingly used by the early 2010s (Hayes 2014). It is not a surprise, then, that affordable retirement and low living costs ranked high for the retirees of both cities as a motive for relocating internationally.

Virtually all retirement destinations across the world include some economic incentive of lower costs of living for those retiring. This has been found, for example, not only in north-to-south migration in the Americas, but also in migration from the United Kingdom to Spain and from Japan to Thailand and Malaysia (van Noorloos 2012; Dixon et al. 2006; Fernández 2011; Hayes 2014; Toyota and Xiang 2012; Warnes 2009; Janoschka 2009). For some, access to affordable health care is another component of this desired, less expensive lifestyle (Pacheco 2016; Janoschka 2009; Kiy and McEnany 2010b).

Although 'affordable retirement' is a motivating theme within both our cities of interest, that descriptor is far more representative of Cuenca than of San Miguel. In the 1970s and 1980s, San Miguel de Allende was considered an affordable retirement location (Covert 2010); however, in the past two decades the trend has been towards increasingly wealthy migrants (Dixon et al. 2006). Cuenca, on the other hand, is still featured heavily as an affordable retirement destination that can buy an upper class lifestyle (Pacheco 2016; Hayes 2018b; International Living 2019b).

This focus on affordability is reflected strongly in our data as well (Table 2.2). As mentioned above, retirees within Cuenca had significantly less monthly income on average than those in San Miguel. These data tie hand in hand with the findings that significantly more retirees in Cuenca cited affordable retirement or health care as the primary motivation for their move when compared to those in San Miguel de Allende, who prioritized lifestyle amenities. Qualitative responses from retirees reflect this as well, with respondents from San Miguel warning others to be aware of the *"increased cost of living which has occurred over the past years"*, whereas those in Cuenca touted it as *"very affordable for someone living on social security and maybe one other retirement plan."*

**Lifestyle Amenities** Just as important as the cost of living are the lifestyle and amenities retirees felt they would gain access to in a new country. For this reason and others, some researchers refer to the international retirement migration movement as 'amenity migration' (van Noorloos 2012; Janoschka 2009; Benson and O'Reilly 2009). There is good reason behind this; research has shown that lifestyle is consistently considered a factor in decisions to relocate in retirement (Pickering et al. 2018). This holds true in Latin America as well; for example, a prior study noted that, among retirees moving to various locales in Mexico, 78.5% considered lifestyle to be a major factor in their decision (Kiy and McEnany 2010b).

"Lifestyle amenities" include climate, an attractive local culture, the opportunity to experience an 'adventure' or to retire in a way atypical of their colleagues, and the pursuit of a different setting or manner of living (van Noorloos 2012; Dixon et al. 2006; Kiy and McEnany 2010b; Pickering et al. 2018; Cuellar Franco et al. n.d.). Retirees often envision their destination cities and countries as places where they can accumulate richer experiences than they would be able to in their home country. As one retiree said of San Miguel de Allende, *"I find SMA rich culturally and intellectually. A great part of it is the large expat community that supports this. I also find that we have a much richer life because of all these activities and all the friendships that are developing through this participation. I am also becoming more and more involved in volunteering to help the community and it is something that adds to the richness of my life..."*

One lifestyle 'amenity' that retirees in our sample considered appealing is the cities' local culture, including their colonial architecture, status as UNESCO World Heritage sites, and the various celebrations and festivals both cities hold. In fact, in both cities, making cultural events a priority was an often-mentioned theme in retirees' suggestions for what could continue to attract other expatriates to the city. The retirees' appreciation of local culture also ties to their sense of a slower, simpler lifestyle. *"In SMA,"* one wrote, *"there are many, many parades and celebrations with fireworks. There seems to be an abundance of joy here and the ability to celebrate life. Traffic is stopped and some things come to a standstill to allow for these events. This would be intolerable in the USA, as it would inconvenience too many people."* This idea of a slower lifestyle, particularly in contrast to their home country, was a common motivating feature often referenced in international retirement migration literature throughout Europe and the Americas (Pickering et al. 2018).

Another significant amenity mentioned often is the virtue of each cities' temperate climate. San Miguel de Allende's is dry and temperate, with cool nights and warm days most of the year, and brief showers in many summer afternoons, providing cooling when it is most desired (Gustafson 2001). Comparatively, Cuenca has been described as the "Eternal Spring" because highs are generally in the 60's or low 70's and residents experience up to 20 rainy days in a month during the rainy season (Hayes 2015). For these reasons, retirees in our sample consistently described the weather in both cities positively, considering Cuenca as "spring every day" and San Miguel de Allende as "fabulous" and "mild". This attention to climate is another recurring theme within international migration literature, described as appealing by expatriates because it often enabled them to escape cold winters and because of the idea that warmer weather facilitates a healthier lifestyle (Pickering et al. 2018).

**Safety and Security** Our survey did not explicitly ask about whether or not safety and security played a factor in retirees' choice of destination location, but this too was brought up often in retiree comments. Safety and security as an appealing factor for retirees has been reflected in other research within North-South migration to Latin America as well (van Noorloos 2012). A focus group conducted with retirees in coastal Mexico commented that they felt safer in Mexico than in the U.S. (Kiy and McEnany 2010b), while another study found that retirees in San Miguel de Allende in particular felt safe and secure compared to the U.S. at the time, in a post-9/11 world (Dixon et al. 2006).

Our sample of retirees consistently mentioned safety within their observations about each city. One, for example, said *"I feel very safe here. I am alert and aware but do walk my dog in the dark, etc. Cuencanos are unfailingly polite and friendly.*

*A great place and our home*." Safety was considered both locally, within retirees' immediate neighborhoods and circles, and on international levels as well. As a San Miguel retiree said, "*San Miguel is a safe place to live and I am so frustrated with the media reports in the US that spin crime here to appear worse than in the US. We feel as safe or more safe in San Miguel than living in the US*."

**Politics** People migrate to Mexico and Ecuador for a variety of reasons. In addition to the 'pull factors' discussed above, 'push factors' can also motivate retirement abroad (López 2017). Politics as a push factor has been a theme in the literature and to a lesser extent in our data. In the Migration Policy Institutes' research, between one-sixth and one-third of retired Americans in Panama and Mexico cited politics as a top factor for their move during their retirement years (Dixon et al. 2006). Reasons included dissatisfaction with the president's political orientation, the general political or social atmosphere, and policies such as high property taxes.

In our data, slightly fewer than 10% of respondents considered 'getting away from home country' as their primary motive for relocating to Mexico or Ecuador. Some respondents reported that during the Barack Obama years conservatives tended to emigrate. As one interviewed expat in Cuenca said, "*there's many Trump Americans here and they mostly came when Obama was president. They were prejudiced, and they didn't want a black president...*" On the other hand, once Donald Trump became president, the political current of emigrating expats shifted to the left. Perhaps ironically, once having immigrated to Mexico or Ecuador, many who came for political reasons develop a more generalized cynical attitude toward the American political system, which they maintain whoever is president.

Despite much feeling about "*the nasty political climate of the U.S.*", expats in both cities maintain interest in U.S. politics. Many cast absentee ballots, and both cities have strong political interest groups. A study in San Miguel de Allende found that 89.4% of retirees continued to vote and cast ballots for U.S. elections even after their move to Mexico (Rojas et al. 2014).

**Government Benefits/Policies** In addition to the traditional 'pull' and 'push' factors discussed above, our survey investigated the importance to retirees of various local or national governmental policies and incentives both in attracting them and in contributing to their quality of life. These are of two general types:

- Policies and programs that exist for all citizens and can be utilized by foreign retirees as well. An example is the INAPAM (Instituto Nacional para las Personas Adultas Mayores) card in Mexico that is given to permanent visa holders and citizens over the age of 60, and which includes discounts on pharmacies, transportation, medical visits, supermarkets, and more (Dixon et al. 2006). These are inclusive enough that immigrant retirees may utilize them as well. Similar benefits, including discounts on airfare, movie theaters, and other activities, can be found in Ecuador for the local elderly (Pacheco 2016).
- Policies and programs that are directly geared towards incentivizing migration. Panama and Costa Rica have a history of visa policies that allow migrants to purchase or import cars every 2 years without paying import duties, and having

**Table 2.3** Perceived importance to immigrant retirees of existing/potential government policies and benefits in Cuenca, Ecuador (n = 424) and San Miguel de Allende, Mexico (n = 325) [a] – Rated on a scale of 1 (least important) to 3 (most important)

| Policy | Mean (SD) Response | | | P-Value for difference between cities[b] |
|---|---|---|---|---|
| | Overall | Cuenca | San Miguel de Allende | |
| **Exemption from:** | | | | |
| Taxes on money brought into the country for purchasing, building, or restoring real estate. | 1.8 (0.9) | 1.7 (0.8) | 1.9 (0.9) | 0.001 |
| Taxes from bringing a car into the country. | 1.5 (0.8) | 1.4 (0.8) | 1.6 (0.8) | 0.110 |
| Paying taxes on social security or other pension income from their home country. | 2.4 (0.8) | 2.5 (0.8) | 2.4 (0.8) | 0.183 |
| Certain municipal taxes, such as selling property | 1.8 (0.9) | 1.6 (0.8) | 2.0 (0.9) | <0.001 |
| **Discounts on:** | | | | |
| Utilities, such as telephone, internet, television, and drinking water. | 2.0 (0.8) | 2.0 (0.8) | 1.9 (0.7) | 0.285 |
| Public transportation. | 2.0 (0.8) | 2.2 (0.8) | 1.8 (0.7) | <0.001 |
| Groceries. | 1.8 (0.8) | 1.9 (0.8) | 1.7 (0.8) | 0.001 |
| Entertainment. | 1.8 (0.7) | 1.8 (0.7) | 1.7 (0.7) | 0.150 |
| **Miscellaneous:** | | | | |
| Maintenance of parks and other recreational opportunities. | 2.3 (0.7) | 2.5 (0.7) | 2.1 (0.8) | <0.001 |
| Provision of free classes in Spanish or opportunities/ "exchanges" where people who are learning Spanish and English can talk to each other. | 2.1 (0.7) | 2.1 (0.7) | 2.0 (0.7) | 0.018 |
| Access to the government health care system after a waiting period. | 2.2 (0.8) | 2.4 (0.8) | 1.9 (0.8) | <0.001 |

[a]Numbers vary due to non-response
[b]Statistic run with Cochran-Mantel-Haenszel

tax exemptions of up to $10,000 on household goods during an initial move (Dixon et al. 2006). Costa Rica's efforts proved so popular that they have since been scaled back due to the influx of migrants the country received (Jackiewicz and Craine 2010).

We asked survey respondents to rate on a scale of 1 (Least Important) to 3 (Most Important) how much they valued certain governmental benefits and incentives. Table 2.3 displays the results. Benefits rated most highly ranged considerably, from being exempt from paying taxes on their monthly income from their home country to the maintenance of parks and other recreational opportunities. The lowest rated incentive was exemption from taxes on bringing a car into the country; perhaps foreseeably, considering the expenses associated with transporting a car to either Mexico or Ecuador.

A few differences emerged between retiree opinions in the two cities. Expatriates in Cuenca prioritized access to affordable health care and discounts over their San Miguel counterparts. Alternatively, residents living in San Miguel de Allende prioritized tax exemptions on purchasing and selling real estate. As will be discussed further in Chap. 3 concerning real estate, retirees in Cuenca are much more likely to rent a house or apartment than retirees in San Miguel, who tend to purchase their homes. This makes the findings in Table 2.3 consistent with the differential income and spending patterns in the two cities.

## 2.2  Engagement of Immigrant Retirees with the Local Community

Survey respondents indicated that the immigrant retiree community in San Miguel de Allende was more established, as the average length of stay for those residents was almost 8 years, compared to retirees in Cuenca who had lived there an average of 4 years. This reflects San Miguel's lengthier history as a general expatriate and retirement, destination. According to data collected between 2004 and 2005, the average American retiree had already been living in San Miguel for 7 years (Rojas et al. 2014). Comparatively, in Cuenca, a number of expatriates that our research team interacted with noted that they felt that retirees new to the city stayed on average for 3 years, after which some decided that Ecuador was not for them in the long-term. This, combined with the fact that the retiree boom in the area is relatively recent, can help explain the differences between cities in terms of longevity of residence.

Although retirees in Cuenca on average may have shorter stays within the country, in other ways the data show more consistent engagement with their new home city. Expatriates in Cuenca were significantly less likely ($p < 0.001$) to make multiple trips back to their country within the past year of data collection. In fact, 37.6% of retirees in Cuenca did not return to their home country once in the previous year and another 43.8% had only visited once. San Miguel immigrant retirees, on the other hand, made more frequent trips home, with the majority traveling back at least twice or more within the previous year (51.3%). These high rates of travel home cannot be solely explained by the potentially easier travel to and from San Miguel to the United States and/or Canada. Another study based in Mexico found that only 20% of American retirees in coastal regions were returning to their home country two or more times a year (Kiy and McEnany 2010a), which speaks to the fact that it is not only Cuenca's lack of an international airport that impacts return trips home.

Another sign of higher integration and commitment to the local community is the difference in banking practices among the retirees in both cities. While the majority of retirees in both locations chose to keep the majority of their savings in their home country (70.7% of retirees within Cuenca, and 79.6% of retirees in San Miguel de Allende), Cuencan retirees were more likely to keep savings either equally between the two countries or completely in their destination country (29.3% versus 14.1%, $p < 0.001$).

One of the principal indicators of integration and engagement into the local community is retirees' ability and desire to speak Spanish. Although this will be discussed in further detail from the perspective of the local residents in Chap. 4, it is worth speaking about the perspectives of the retirees on the necessity (or not) of learning the Spanish language in order to live comfortably in each city. Language has consistently been a point of difficulty in terms of integration and lifestyle for expatriates immigrating to countries in which their native language isn't spoken (Rojas et al. 2014; Hayes 2018a; Casado-Diaz 1999; Croucher 2011; Gustafson and Laksfoss Cardozo 2017), and was spoken about often within our sample.

Among our survey respondents, most retirees in both cities reported some grasp of Spanish, although very few would be described as proficient. 43.5% described their ability as able to hold simple conversations, and another third of the respondents attested they were able to speak 'limited' Spanish. One retiree spoke to a potential reason for the lack of proficiency, "*language – it's hard for some older people to learn Spanish*". However, she then added the importance of doing so: "*but not being able to communicate with locals limits your understanding of what's going on.*" Although retirees were fairly limited in their current language ability, they showed a desire to advance their skills. 59.6% of retirees wanted to improve to the level of being able to have extensive conversations easily, while another 24.1% wanted the ability to conduct business transactions easily.

Responses showed that a number of retirees felt strongly about the impact that learning the language would have on one's ability to integrate smoothly into the community, another theme reflected in prior interviews with retirees in Cuenca (Hayes 2015). One expatriate in San Miguel attributed some points of tension between the immigrant community and local community to the inability to communicate: "*I have seen gringos misinterpret things in a serious way, e.g. thinking a healthcare professional has cheated them, because they do not speak/understand Spanish. Anyone who lives here should study the language seriously and continually for their own benefit and for thanks and respect to Mexico*". This coupling of language and 'respect' for the city was a theme throughout our survey and interview responses; a number of retirees expressed the feeling that people who made no attempt to learn the local language were also being dismissive of the local culture.

## 2.3 Engagement of Retirees with Others in the Expatriate Community

Another aspect of social life for retired immigrants involves interaction with, and integration into, existing expatriate communities in each city, both of which are now sizeable enough to have a fairly large social network for newcomers. Expatriates interact and network in a number of ways, including online blogs, philanthropy events, political groups, and 'gringo' social events (Kiy and McEnany 2010a; Pacheco 2016; Croucher 2011; Nudrali and O'Reilly 2016).

In the last few years, expatriates interested in both cities have found a relatively ready and willing community available to engage with early in the decision to relocate, which embraces them socially once they arrive. These circles often begin online, with resources like *Gringo Tree* and *Gringo Post* specific to Cuenca, or *Yahoo Listservs*, like the ones found in San Miguel de Allende (Pacheco 2016), which then transition to in-person relationships once an expatriate visits or relocates.

Access to a thriving social network of people with a similar lifestyle is a motivating factor for many retirees to choose certain cities (Pickering et al. 2018). As one of our survey respondents wrote, *"expats in SMA [San Miguel de Allende] are a self selecting group. Everyone is desirous to help others acclimate and thrive in the city. It is easy to make social connections in a way different than in the US..."*, or as another commented, *"this location [Cuenca] attracts highly educated creative people. Expats see each other as a fellow adventurer."* There is a sense from the surveyed retirees that many feel they have found a network of people who share similar values, interests, and character traits. As in the example above, 'being adventurous', is one such trait that many retirees attribute to themselves, as was reflected in their decision to move abroad (Cuellar Franco et al. n.d.).

Some events in which retirees interact are specifically meant to cater to foreign retirees (Kiy and McEnany 2010a; Clausen and Velázquez García 2018). This occurs in retirement destinations outside of Latin America as well – for example, football bar nights for British expats in Didim, Turkey (Nudrali and O'Reilly 2016). Through fieldwork, our research team saw evidence of the existence of similar phenomena in both cities, with popular café locations advertising in English and hosting 'Gringo Nights' geared toward retirees (see, as an example, Fig. 2.2).

**Fig. 2.2** A research assistant sits with a retired expatriate within Cuenca, Ecuador. The setting, the Sunrise Café, is a popular gathering place for retirees. As can be seen in the background of this photo, a sign advertises a weekly mini-fair in English and Spanish

Philanthropy and non-profit volunteering is another activity that has a heavy expatriate presence and social component. Philanthropy within retirement destinations has tended to be an avenue for retirees to integrate themselves into the community, as well as integrate into expatriate social circles (Croucher 2011; Casado-Diaz 2016; Foulds 2014). Our sample showed high levels of involvement in giving back to the community in various ways – these included volunteering and/or working with local organizations and non-profits, giving contributions to various causes (52.7% of respondents), and directly donating to local families (49.8%). A number of respondents specifically pointed out that the non-profit organizations they were involved in were 'gringo' led. Other contributions to the local community identified by survey respondents were paying property taxes, starting new businesses, and otherwise participating in the local economy.

This is not to say that there were not problems uncovered among the expatriate circles. The transitory nature of many migrants can lead to difficulty in nurturing friendships. As one retiree in Cuenca commented, *"There's a level of fluidity – lack of commitment – to living in the expat community that is Cuenca. People come and go all the time for various lengths of time for various reasons. Few are those who are in Cuenca…"stay put"and develop relationships, friendships, that can deepen and be satisfying over time."* This theme was more prevalent in Cuenca, which is reflective of the shorter stay of retirees.

## 2.4 Immigrant Retiree Opinions' on Their Impact Within the Local Community

In a series of open-ended questions, respondents were asked to consider the overarching phenomenon of international retirement migration. As part of the research's goal to understand the impact of retirement migration on host communities, we asked respondents to share their perceptions of potential negative impacts that might result from international retirement migration to their new home city. A number of clear themes and patterns became evident in responses across both cities, including concerns that their presence may raise the overall cost of living, change the local culture, have a negative impact on the environment, and contribute to negative stereotypes of Americans.

### 2.4.1 Concerns About Raising the Cost of Living and Housing

The impact most often mentioned in both cities was contributing to a general rise in the cost of living for the local population, with nearly half of retirees who wrote in a response to our open-ended question about negative impact specifically mentioning this possibility. Although retirees saw positive aspects to their presence in the form of economic stimulation and increased job opportunities, many respondents acknowledged that unintended negative consequences could potentially accompany those positive impacts. As one respondent from Cuenca said, negative impacts can stem from *"using origin-country economic conventions without considering their impact on locals, such as tipping or over-paying for goods/services because you can. These lead to competition for gringo business which negatively impacts locals and can leave them at the back of the line."* Another respondent, from San Miguel de Allende, acknowledged the fact that these norms are not due to mal-intent and in fact come from a desire to help, but *"often what happens however, admittedly with the best of intentions, is the expats are responsible for bidding up prices and wages in the parts of the economy they impact, beyond what the rest of the community can sustain."* These opinions and others acknowledged the economic power that retirees bring with them, and how this power, combined with differences in cultural norms between the retirees and the communities within host cities (i.e. the tendency to tip or pay hourly versus daily wages), can have unintended negative consequences.

A term used almost exclusively by respondents in Cuenca to explain rising living costs was "gringo pricing". This referred to purposive 'gouging' on the behalf of the locals towards the expat community. As one expat said, *"locals can believe that all expats are 'rich' and Gringo price us all"*. This sense of a "racialized price system" was identified by migrants in another study conducted in Cuenca (Hayes 2015), in which, during interviews, retired migrants often said that they were targeted due to their obvious stature as white, most likely American, retirees. Retirees in San Miguel, as mentioned above, were certainly cognizant of rising prices but generally attributed them to the retired immigrant community, wealthy Mexican nationals from Mexico City, and tourists.

A related issue raised by many retiree survey respondents and interviewees was whether, and if so, to what extent their presence contributed to increases in housing prices. Retirees acknowledged that apartment and home owners had been able to raise their rental and sale prices because retired expatriates could pay them, whereas many in the local population could not. Retiree respondents also agreed that real estate agents would prefer to rent to gringos, saying *"rising prices, especially housing rentals, because Ecuadorian landlords prefer gringos, thinking they care for the properties better"*. In San Miguel de Allende, respondents also referred to displacement of households from the center of the city due to rising real estates prices. One retiree commented that, *"property prices are rising because retirees and others are willing and able to pay much more than most locals can afford. Now many Mexican youth won't be able to buy a home in the town where their families have lived for many generations."*

## 2.4.2 Concerns About Changing the Local Culture

Another commonly mentioned concern related to retiree impact on the culture of each city – that the expatriates might be exporting the culture of their home countries to their Latin American host cities to the expense of the pre-existing local culture. Approximately a third of retirees who wrote in response to our open-ended question about negative impact referenced worries about loss or change of the local culture due to increasing migration. One said, "*I came here to live within a new culture. If there were any kind of mass migration of my countrymen and women, that would become less pure and more difficult to do.*" Others echoed this sentiment by lamenting the loss of small-town or community feel of the cities due to higher numbers of retirees and other visitors.

Some retirees specifically spoke of a perceived Americanization of the city in the form of the adaptation of American holidays and customs, such as Halloween celebrations rather than Day of the Dead, and a proliferation of American stores. One respondent regarded the "*increased emphasis on US culture – stores, shops, restaurants catering to tourists, rather than natives*" and the "*intrusion of US franchises, such as McDonalds, Burger King, Costco, Walmart, Home Depot, Radio Shack, Gold's Gym, etc.*" as a general shortcoming of retiree migration to the community. The "Americanization" of the cities was perceived to be spearheaded by a subset of Americans who "*try to create a 'mini-America' around themselves and don't associate with the native community*".

In both cities, albeit more so in Cuenca than San Miguel, retirees expressed concern over a small subset of "obnoxious retirees", otherwise known as "ugly Americans" (Hayes 2015). One respondent complained of this group, saying the problem lies with "*retirees who act like the 'Ugly American' constantly criticizing or demanding Cuenca change to their liking.*" This group was thought to consider their home countries to be superior in many ways and was described as trying to change local culture to fit the values and lifestyle of North America. One respondent, for example, described such retirees as "*people who, though maybe well intentioned, try to make changes in the culture that they think would 'make things better'.*" Another retiree spoke to "*demands that services become 'more like back home' negating some of the unique, charming native customs.*" Other complaints by respondents reflected the sense that this minority of "obnoxious retirees" is not respectful of the natives or the culture of the community to which they have relocated. They are thought to have chosen to live in these communities only because they can live more luxuriously and have no consideration or appreciation of the value of the people, architecture, events, and customs that make up the local culture. Respondents often distanced themselves as much as possible when referring to this group and attributed a number of the negative impacts of retirement migration to the "obnoxious retirees". Another respondent agreed, expressing concern that social dynamics between the retiree group and the local community were becoming strained through rude, abrasive behavior in public settings.

### 2.4.3  Belief That Migration Has Little or No Impact on Host Communities

Surprisingly few respondents stated that they felt there were no potentially negative changes happening within the communities due to the presence of migration and retired immigrants. All but one who responded this way were from Cuenca; they felt that the low number of retirees in relation to the city's population meant that their impact would be minimal. One such respondent hypothesized about negative impacts, but followed up to say, "*this is a big enough city that any of the effects I've named would only result from a sudden, really serious inundation of expats...like tens of thousands in a year, say...*".

In San Miguel de Allende, on the other hand, a minority of respondents identified negative trends but attributed them either to a combination of tourism, retirement migration, and second-home ownership by wealthy Mexicans from other cities or solely to these other, outside forces. As one wrote, "*the negative impact on San Miguel is no longer the retirement migration. It is the unfettered construction of housing for tourists – Mexican tourists...or weekend visitors.*"

### 2.4.4 Differences Between Retirees in San Miguel de Allende Versus Cuenca Regarding Perceived Impacts on Each City

While the above issues were major topics of concern in both cities, there were a few differences between respondents in San Miguel de Allende and Cuenca regarding perceived retiree impact. Most prominent was attitudes regarding impact of retirement migration on the environment. In San Miguel de Allende, a significant number of retirees voiced concerns over negative impact on the urban environment, particularly in terms of water quality and availability, and the increasing presence of traffic within the city center. These apprehensions were mentioned solely in San Miguel; virtually no respondents in Cuenca referenced environmental issues due to international retirement migration. San Miguel de Allende's struggle with water quality and availability became a tangential theme of the research, not just within survey answers from retired expatriates, but in interviews with local residents as well. As one retiree wrote, "*water is becoming a critical issue in terms of both quality (increasing levels of fluoride and arsenic in our deep wells and general availability). This is a significant problem for the many smaller communities around SMA*". However, retirees in San Miguel did not necessarily attribute these problems to foreign retirees within the city; instead, respondents spoke of the general population surge and overdevelopment that the city has experienced. Themes regarding the increasing traffic within the city center were similar. Although some retirees attributed this to retirees "*bringing their automobiles with them*", others spoke about traffic more generally and related the rise in congestion to tourists and weekend visitors. Rising traffic has been on the municipality of San Miguel's mind as well; in a published paper the city outlined a number of infrastructure and road constructions proposed to help improve traffic flow (Ayuntamiento de San Miguel de Allende 2017).

## 2.5 Summary and Conclusion

Through our online survey with retired immigrants living in San Miguel de Allende, Mexico and Cuenca, Ecuador, and our on-site field work at each location, our research team was able to connect with hundreds of retired immigrants in both cities. Our findings proved to largely confirm those of other studies investigating retired immigrants in Latin America (Kiy and McEnany 2010a; Rojas et al. 2014; Álvarez et al. 2017), while adding texture by the volume of respondents, the types of questions, and the comparison of the two cities.

It is clear that "immigrant retirees" or "expat retirees" are by no means a homogenous group. Yes, there are certain patterns and themes among responses to our online surveys and in our field notes, but our research also shows that the retirees represent a variety of backgrounds, motivations, activities, values, and opinions. The majority came from the United States, but even within this pool, respondents ranged in their demographic characteristics, their reasons behind moving, their engagement with the local community, and their perspectives on retirement migration and its impact.

The differing backgrounds of the retirees affected the ways they learned about and pursued international retirement, the factors that ultimately guided their decisions to move, and the ways in which they eventually navigated and integrated into each city after relocation. This fact is particularly highlighted by the significant differences found virtually across the board between the immigrant retiree communities in San Miguel de Allende and Cuenca, in terms of demographic characteristics, motivations for moving, the relocation decision-making process, their levels of local community engagement, and their perspectives on the impact of retirement migration on the local community.

While the majority of respondents move for an affordable retirement and/or various lifestyle amenities in their chosen destination (van Noorloos 2012; Dixon et al. 2006; Fernández 2011; Pacheco 2016; Hayes 2014; Toyota and Xiang 2012; Janoschka 2009; Kiy and McEnany 2010b; Pickering et al. 2018; Cuellar Franco et al. n.d.), respondents in our two study cities differed in terms of the factors that most attracted them to their chosen location. In general, retirees to San Miguel de Allende were older, wealthier, and more concerned with improving their quality of life through lifestyle amenities, where as those who had retired to Cuenca more often described a move motivated by economic hardships they would face in their home country, findings that are consistent with prior research on the retiree communities in these cities (Dixon et al. 2006; Rojas et al. 2014; Álvarez et al. 2017; Pacheco 2016; Covert 2010; Hayes 2018b).

Our survey indicated that many expat retirees were concerned about their potential impact on the local community. Respondents were particularly mindful of the impact that they and others may have had on the cost of living within their retirement city and on the possibility that customs and adaptations they brought with them may have changed the fabric of the local community. They were also cognizant of other potential effects of retiree migration, including environmental ramifications and impact community infrastructure and social networks. Respondents spoke of the importance of mitigating these potential negative impacts as much as possible, but it is unclear how many retirees take actions to reduce these impacts or what those actions may be. Some cited making an effort to communicate with and participate within the local community and contribute through volunteering and philanthropy work. Other respondents, however, tended to downplay or distance themselves from these issues. Some attempted to remove themselves entirely from responsibility, attributing negative impacts solely to a small subset of retirees referred to as the "obnoxious retiree" or the "Ugly American"; others placed the responsibility on tourists or other immigrant groups.

Ultimately, the picture of retirees that has emerged from our data is one of a nuanced group, reflecting a large array of beliefs and thoughts regarding their contributions to, impact on, and interaction with the Mexican or Ecuadorian community they have decided to make home.

# References

Álvarez, M. G., Guerrero, P. O., & Herrera, L. P. (2017). Estudio sobre los impactos socio-económicos en Cuenca de la migración residencial de norteamericanos y europeos: Aportes para una convivencia armónica local. In *Informe Final*. Cuenca: Avance Consultora.

Ayunamiento de San Miguel de Allende. (2017). *Avanzamos Segundo a Segundo: Informe de Resultados*. San Miguel de Allende, Mexico: H. Ayuntamiento 2015–2018, San Miguel de Allende. https://sanmigueldeallende.gob.mx/transpa_prueba/docs/15/15061127331.pdf. Accessed 8 July 2019.

Bantman-Masum, E. (2015). Migration machine: Marketing Mexico in the age of ICTs. In O. Frayssé & M. O'Neil (Eds.), *Digital labour and prosumer capitalism: The US matrix*. Houndmills/Basingstoke/Hampshire: Palgrave Macmillan.

Benson, M. C., & O'Reilly, K. (2009). Migration and the search for a better way of life: A critical exploration of lifestyle migration. *Sociology Review, 57*(4), 608–625.

Casado-Diaz, M. A. (1999). Socio-demographic impacts of residential tourism: A case study of Torrevieja, Spain. *International Journal of Tourism Research., 1*(4), 223–237.

Casado-Diaz, M. A. (2016). Social capital in the Sun: Bonding and bridging social capital among British retirees. In M. Benson & K. O'Reilly (Eds.), *Lifestyle migration: Expectations, aspirations and experiences* (pp. 87–102). New York City/Oxon: Routledge/Taylor & Francis Group.

Clausen, H. B., & Velázquez García, M. A. (2018). National Mexican tourism policy and north American second home owners in Mexico: Local tourism development and Mexican identity. In M. Hall & D. K. Müller (Eds.), *The Routledge handbook of second home tourism and mobilities* (pp. 64–74). New York City/Oxon: Routledge/Taylor & Francis Group.

Covert, L. P. (2010). Defining a place, defining a nation: San Miguel de Allende through Mexican and foreign eyes. In *Dissertation for a degree of doctor of philosophy within Yale University*. Ann Arbor: Proquest.

Croucher, S. L. (2011). *The other side of the fence: American migrants in Mexico*. Austin: University of Texas Press.

Cuellar Franco, J. L., Hidalgo, M. G. A., Sánchez, J. C. M. (n.d.). Allende. *Enciclopedia de los Municipios y Delegaciones de Mexico, Ayuntamiento de Allende*. http://siglo.inafed.gob.mx/enciclopedia/EMM11guanajuato/municipios/11003a.html. Accessed 10 July 2019.

Dixon, D., Murray, J., & Gelatt, J. (2006). *America's emigrants: U.S. retirement migration to Mexico and Panama*. Washington, DC: Migration Policy Institute.

Emling, S. (2010, May 18). Americans who seek out retirement homes overseas. *The New York Times*. https://www.nytimes.com/2010/05/19/your-money/19iht-nwmove.html. Accessed 13 June 2019.

Fernández, I. G. (2011). The right to the city as a conceptual framework to study the impact of north-south migration. *Recreation and Society in Africa, Asia and Latin America., 2*(1), 3–33.

Foulds A. (2014). Buying a colonial dream: The role of lifestyle migrants in the gentrification of the historic center of Granada, Nicaragua. *Theses and Dissertation – Geography, 18*. https://uknowledge.uky.edu/geography_etds/18/. Lexington: University of Kentucky.

Gustafson, P. (2001). Retirement migration and transnational lifestyles. *Ageing and society., 21*(4), 371–394.

Gustafson, P., & Laksfoss Cardozo, A. E. (2017). Language use and social inclusion in international retirement migration. *Social Inclusion, 5*(4), 69–77.

Hayes, M. (2014). 'We gained a lot over what we would have had': The geographic arbitrage of America's lifestyle migrants to Cuenca, Ecuador. *Journal of Ethnic and Migration Studies., 40*, 1953–1971. https://doi.org/10.1080/1369183X.2014.880335.

Hayes, M. (2015). Moving south: The economic motives and structural context of North America's emigrants in Cuenca, Ecuador. *Mobilities, 10*(2), 267–284.

Hayes, M. (2018a). The gringos of Cuenca: How retirement migrants perceive their impact on lower income communities. *Royal Geographical Society: Area Journal, 50*(4), 1–9.

Hayes, M. (2018b). *Gringolandia: Lifestyle migration under late capitalism*. Minneapolis and London: University of Minnesota Press.

International Living. (2019a). World rankings: The best places to live. *International Living*. https://internationalliving.com/world-rankings/. Accessed 28 June 2019.

International Living. (2019b). Cuenca, Ecuador: Cuenca is famous for its colorful festivals, distinct food, and breathtaking scenery. *International Living*. https://internationalliving.com/countries/ecuador/cuenca/. Accessed 19 June 2019.

Jackiewicz, E. L., & Craine, J. (2010). Destination Panama: An examination of the migration-tourism-foreign investment nexus. *RASALA, 1*(1), 5–29.

Janoschka, M. (2009). The contested spaces of lifestyle motilities: Regime analysis as a tool to study political claims in Latin American retirement destinations. *Die Erde: Special Issue "Amenity Migration", 140*(3), 1–20.

Kiy, R., & McEnany, A. (2010a). *U.S. retirement trends in Mexican coastal communities lifestyle priorities and demographics*. International Community Foundation.

Kiy, R., & McEnany, A. (2010b). *Housing and real estate trends among Americans retiring in Mexico's coastal communities*. International Community Foundation.

Li Ng, J. J., & Ordaz Díaz, J. L. (2012). US baby boomers in Mexico: A growing group of immigrants. *BBVA Research*. https://www.bbvaresearch.com/KETD/fbin/mult/120321_MigracionMexico_09_eng_tcm348-378589.pdf. Accessed 26 July 2019.

Lizárraga-Morales, O. (2009). Networks and social impact of residential tourism for US retirees in Mazatlan, Sinaloa & Cabo San Lucas, Baja California Sur. *Journal of Architecture, Urbanism and Social Sciences., 1*(3), 1–10.

López, E. (2017). Estudio y Análisis para la creación de un "Programa de Capacitación en Idioma Inglés para Conductores de Taxi en la cuidad de Cuenca". In *Dissertation for a degree from the Universidad del Azuay, Escuela de Estudios Internacionales*. Cuenca: Universidad del Azuay.

Nudrali, O., & O'Reilly, K. (2016). Taking the risk: The British in Didim, Turkey. In M. Benson & K. O'Reilly (Eds.), *Lifestyle migration: Expectations, aspirations and experiences* (pp. 137–152). New York City/Oxon: Routledge/Taylor & Francis Group.

Ocampo, N. (2019). Enjoying an active life in Costa Rica. *International Living*. https://internationalliving.com/enjoying-an-active-life-in-costa-rica-rrei/. Accessed 10 July 2019.

Opdyke, J. D. (2008, February 5). Retiring abroad may not be paradise. *The Wall Street Journal*. https://www.wsj.com/articles/SB120215009626041505. Accessed 10 July 2019.

Otero, L. M. Y. (1997). U.S. retired persons in Mexico. *American Behavioral Scientist, 40*(7), 914–922.

Pickering, J., Crooks, V. A., Snyder, J., & Morgan, J. (2018). What is known about the factors motivating short-term international retirement migration? A scoping review. *Population Ageing*,
1–17.

Pacheco, P. S. V. (2016). Analysis of the impact of digital communication media which promotes
Cuenca as a residential destination for American retirees, over the period 2010–2015. In
*Dissertation for a degree from the Universidad del Azuay, School of International Studies*.
Cuenca: Universidad del Azuay.

Rojas, V., LeBlanc, H. P., & Sunil, T. S. (2014). US retirement migration to Mexico: Understanding
issues of adaptation, networking, and social integration. *Journal of International Migration &
Integration, 15*(2), 257–273.

Schafran, A., & Monkkonen, P. (2011). Beyond Chapala and Cancún: Grappling with the impact of
American migration to Mexico. *Migraciones Internacionales, 6*(2), 223–258.

Sheridan, M. B. (2019, May 18). The little-noticed surge across the U.S.-Mexico border: It's
Americans, heading south. *Washington Post*. https://www.washingtonpost.com/world/the_
americas/the-little-noticed-surge-across-the-us-mexico-border-its-americans-heading-south/2
019/05/18/7988421e-6c28-11e9-bbe7-1c798fb80536_story.html?utm_term=.194b5e0a0a14.
Accessed 10 July 2019.

Social Security Administration. (2000). *Annual statistical supplement to the social security bulletin, 1999*. Washington, DC: United States Congress, Office of Retirement and Disability Policy,
Office of Research, Evaluation, and Statistics.

Social Security Administration. (2019). *Annual statistical supplement to the social security bulletin, Dec 2017*. Washington, DC: United States Congress, Office of Retirement and Disability
Policy, Office of Research, Evaluation, and Statistics.

Sunil, T. S., Rojas, V., & Bradley, D. E. (2007). United States' international retirement migration:
The reasons for retiring to the environs of Lake Chapala, Mexico. *Ageing & Society, 27*(4),
489–510.

Talty, A. (2017, October 18). Baby boomers are retiring abroad in droves... But is it right for you?
*Forbes*. https://www.forbes.com/sites/alexandratalty/2017/10/18/baby-boomers-are-retiring-
abroad-in-droves-but-is-it-right-for-you/#1469efeb5f7b. Accessed 10 July 2019.

Toyota, M., & Xiang, B. (2012). The emerging transnational "retirement industry" in Southeast
Asia. *International Journal of Sociology and Social Policy., 32*(11/12), 708–719.

Truly, D. (2002). International retirement migration and tourism along the Lake Chapala Riviera:
Developing a matrix of retirement migration behavior. *Tourism Geographies: International
Journalism of Tourism Space, Place, and Environment, 4*(3), 261–280.

van Noorloos, F. (2012). *Whose place in the sun? Residential tourism and its implications for
equitable and sustainable development in Guanacaste, Costa Rica*. Delft: Uitgeverij Eburon.

van Noorloos, F. K., & Steel, G. (2016). Lifestyle migration and socio-spatial segregation
in the urban(izing) landscapes of Cuenca (Ecuador) and Guanacaste (Costa Rica). *Habitat
International., 54*, 50–57.

Warnes, T. A. (2009). International retirement migration. In P. Uhlenberg (Ed.), *International
handbook of population aging*. New York/Berlin/Heidelberg/Dordrecht: Springer.

# Chapter 3
# Real Estate, Housing, and the Impact of Retirement Migration in Cuenca, Ecuador and San Miguel de Allende, Mexico

**Philip D. Sloane and Johanna Silbersack**

The relationships around housing and retirement migration, particularly within historic, colonial cities in Latin America, are complex. A differential in the cost of housing between the origin and destination country is clearly one of many motivating factors in international retirement migration (Bernier 2003); however, how this migration impacts housing prices and availability on the local level has not been extensively studied in the retirement migration literature, although it is a commonly referenced concern.

Real estate is one of the key components of the 'migration industry', or 'country marketization' process of international retirement migration. Real estate contributes to the industry in a number of ways (Pallares and Rollins-Castillo 2019; David et al. 2015; Rainer 2019). By relaxing limits on foreign ownership or advertising low property taxes, real estate policy can be used by governments as a tool to incentivize migration. Additionally, the industry itself is a lucrative one. In Malta, for example, there are real estate agencies that specialize solely in foreign buyers and advertise the lifestyle amenities of the island, including luxury living (David et al. 2015).

Real estate agencies operating in international retirement destination communities are not necessarily started in or led by the local residents; indeed, many North American real estate companies have expanded their reach into retirement destinations for this purpose – including Century 21, Coldwell Banker, RE/Max, and others (Rainer 2019). Whether and if so to what extent these companies and this industry raise prices across the board on home purchases and rentals in retirement destinations is unclear, although some findings suggest that they have (Rainer 2019; van Noorloos and Steel 2016; Bastos 2014).

In established cities with historic buildings, such as those in Latin America that have achieved UNESCO World Heritage designation, rising real estate costs come

P. D. Sloane (✉) · J. Silbersack
University of North Carolina at Chapel Hill, Chapel Hill, NC, USA
e-mail: philip_sloane@med.unc.edu

© Springer Nature Switzerland AG 2020     41
P. D. Sloane et al. (eds.), *Retirement Migration from the U.S. to Latin American Colonial Cities*, International Perspectives on Aging 27,
https://doi.org/10.1007/978-3-030-33543-4_3

hand in hand with gentrification, a process in which the purchase and renovation of historic buildings markedly increases housing prices, often driving out the original residents (Janoschka et al. 2014). This process, which has been termed "re-staging," involves renovation of historic colonial architectural buildings and facades and their conversion into luxury housing, residential estates, boutiques, and hotels, while existing street vendors and communities are pushed out and removed (Janoschka et al. 2014; Foulds 2014; Steel and Klaufus 2010). It has been occurring in many Latin American World Heritage cities, including Antigua, Guatemala; Granada, Nicaragua; Cuenca, Ecuador; and San Miguel de Allende, Mexico – the latter two being the focus of this chapter and book (Janoschka et al. 2014; Pacheco and Vallejo 2016; Navarrete 2018).

Typically, the greatest impact of retirees is on relatively high-end housing, since their incomes generally place them among the wealthiest in the local economic order; however, at times retirees may compete with local residents for less expensive housing as well (Casado-Diaz 1999). Part-time residence and its impact on housing markets is another issue, as is the question of whether and how often second homes are purchased and later converted to more-or-less permanent use (Williams et al. 1997).

Cuenca and San Miguel de Allende are no strangers to this process. Researchers found that within Cuenca's historic buildings, local residents have been moving to make way for people who can afford a costlier living style, as entertainment and restaurants became more prevalent in the area (van Noorloos and Steel 2016). Within San Miguel de Allende, there have been clear signs of gentrification in the city center, along with the growth of foreign communities with higher purchasing power (Navarrete 2018; Schafran and Monkkonen 2011). Some researchers have estimated that as many as 85% of the homes in the city center are now owned by persons who are not natives of the city (Lizárraga-Morales 2008) and that immigrants from the United States and Canada make up 60% of residents within the historic center (Navarrete 2018).

In our research in Cuenca, Ecuador and San Miguel de Allende, Mexico, we wanted to further investigate the impact of immigrant retirees on the real estate market through the lens of the local population. Our research questions included:

- What housing choices are made by retirees moving to these cities, and do the housing patterns differ between the cities?
- Had the influx of migrant retirees affected housing prices, and if so, did changes in housing price affect the cost of living for local residents?
- Given that both cities had UNESCO heritage status because of the charm and beauty of their central areas, would the migrant retirees concentrate in historic areas, causing gentrification and relocation of local residents to peripheral neighborhoods?
- What were the attitudes of the local residents and retired migrants on the impact of retirees on the real estate market? Did their opinions concur?
- What other factors are important in understanding the impact of migrant retirees on real estate in these cities?

To address these questions, data were drawn from the following sources: Interviews (in Spanish) with real estate agents and a sample of other local residents in the two study cities; an internet survey (in English) of expat retirees aged 55 and older living in the study cities; statistical data on housing costs (Ecuador only), an internet inventory of properties for sale in October, 2018; data collector field notes, and photographs. Details on the study's data collection methods are provided in the Appendix.

## 3.1 Legal Requirements for Americans to Purchase Real Estate in Cuenca and San Miguel de Allende

For retirees who are thinking of moving abroad, a major consideration is whether they are interested in buying a home in their new location. Many will rent initially, but for long-term residence, it often makes sense to purchase and own a home, especially if home ownership in the new country appears to be a good investment. Both Ecuador and Mexico have requirements and restrictions regarding the purchase of real estate by foreigners, however; and so retirees must accommodate to the different regulations that govern home ownership by foreigners.

Historically, Mexico has not allowed foreigners to purchase land in any capacity. In fact, Mexico's Constitution in 1917 established that only Mexicans by birth or naturalization could own any land, effectively restricting foreign access to property (United Mexican States Constitution 2019). Since then, these laws have been greatly relaxed. By 1973, the Foreign Investment Law was passed, revising the Constitutions' Article 27, Section 1 to allow foreigners to own land throughout Mexico, with the exception of a 'Restricted Zone', defined as land within 50 km of the coast or 100 km of the border (Schafran and Monkkonen 2011; Global Property Guide 2017). Today, it is possible for foreigners to purchase land even within the restricted areas through a few mechanisms put into place in the early 1990s (Smith et al. 2009), which allow purchase either by way of a bank 'trust' (fideicomiso) or by registering a company with a Mexican address (Schafran and Monkkonen 2011; Lizárraga-Morales 2010).

Regardless of whether foreign retirees are looking for land within or outside of the 'Restricted Zone', they face additional real estate regulations compared to citizens. Any purchase made by foreigners in Mexico must be formally approved by the Secretaria de Relaciones Exteriores before being made legal. To receive approval, retirees fill out a FF-SRE-006 form, which requires documentation proving their legal stay in Mexico, descriptions and measurements of their desired property, and formally waives their "right to invoke the protection of [their] government" in regard to any asset they purchase (Secretaria de Economica 2016).

In contrast, Ecuador's property laws do not distinguish between citizens or foreigners at all (Bayer 2018). In fact, purchasing property in Ecuador can help make a retiree eligible for an 'Investor Visa', a type of extended stay visa which requires either an investment of $30,000 or more into a Ecuadorian business, placement of at least $26,250 in an Ecuadorian bank, or purchase of property worth at least $30,000 (Bayer 2018; Haines 2018a).

This does not, however, imply that the purchasing process is simple for foreigners in Ecuador. Language barriers and differences in the real estate market, combined with multiple steps of bureaucracy, can create challenges for retirees looking to purchase property in Cuenca (Bayer 2018; Haines 2018b). Local residents often bypass realtors altogether and list homes themselves; the real estate market lacks 'comparatives' for potential buyers to review; and multiple listing services are not nearly as available as in the United States or Canada (Bayer 2018; Haines 2018a, b, c). Also important for retirees to consider is access to financing. The real estate market in Ecuador is primarily a cash market with limited options for mortgages, and foreigners can have an even more difficult time accessing mortgages due to restrictive bank policies (Bayer 2018; Chaca 2018; Gregor 2014).

The purchase process in Ecuador also requires that potential buyers navigate multiple levels of bureaucracy. Buyers are required to verify there are no zoning issues through "Línea de Fábrica", to notarize an agreement of purchase known as 'Promesa de Compraventa', and to notarize and submit to the "Registro de la Propiedad" a new title. Having an interpreter fluent in Spanish and familiar with local government is helpful in negotiating all of these steps (Bayer 2018; Registros n.d.; Municipal de Cuenca n.d.).

## 3.2 Housing Patterns of Migrant Retirees Live in Cuenca and San Miguel de Allende

For migrant retirees, Cuenca tends to be a city of renters and San Miguel de Allende a city of owners (Fig. 3.1). In Cuenca, of the 370 retired expats who responded to our survey, the vast majority (73%) lived in a rental unit, with half of all retired expats living in rental apartments. The situation in San Miguel de Allende is quite different; of the 275 respondents to our online survey who designated their housing, 59% lived in a house that they owned and another 14% lived in a condominium apartment that they owned.

Why do retired immigrants tend to rent in Cuenca and to buy in San Miguel de Allende? One explanation could be the economic status of the migrants, in that those who have moved to Cuenca by and large have fewer resources. Our survey supported this, in finding that retirees in Cuenca self-described themselves as 'economic refugees', with 63% of retiree respondents from Cuenca reporting a monthly incomes of less than $3000 a month; comparatively, 58% of San Miguel's retirees reported a monthly income of over $3000. Uncertainty about the economic long-run is another, especially since Cuenca is relatively new as a host city for retirees and is closer to countries like Colombia and Venezuela. As one retiree in Cuenca told us, *"In Latin America, you are only one president away from a Venezuela."*

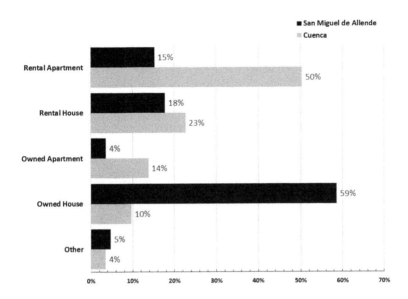

**Fig. 3.1** Relative proportions of immigrant retirees who rent versus own their homes in Cuenca, Ecuador and San Miguel de Allende, Mexico. Differences between cities are statistically significant at $p < 0.001$

Another reason for this differential in purchasing versus renting is the real estate market itself, in that the rapid rise in property prices and the tourist boom in San Miguel de Allende make buying a home appear to be a good investment, even (or especially) for individuals who plan to live in the home for only part of the year. In contrast, Americans in Cuenca are more likely to regard other elements of retirement migration as more attractive.

To help document and compare the housing patterns of expat retirees and the locals, we asked all expats aged 55 or older who completed our online surveys and all locals we interviewed to place their home on a map of their city. We also asked real estate agents and other native-born residents we interviewed to assign self-reported neighborhoods with a number from 0 to 3 to indicate the prevalence of expat retirees in that neighborhood. Using these data, we constructed a series of maps.

We combined the data from our informants in each city to create density cloud maps of Cuenca and San Miguel de Allende, which demonstrated the prevalence of gringos by neighborhood. This method has been elegantly shown in a previous study in Cuenca to generate similar results to data from property tax records or from surveys of retirees (Álvarez et al. 2017). We used the 510 residence locations provided by our online survey respondents (300 in Cuenca and 210 in San Miguel de Allende), and the residency locations from the native residents we interviewed to separately map their locations, providing us another method of plotting where people live in these two cities.

Cuenca. Figure 3.2 is a density cloud map of Cuenca, indicating where in the city immigrant retirees, most of whom are Americans, tend to live. Neighborhood A on

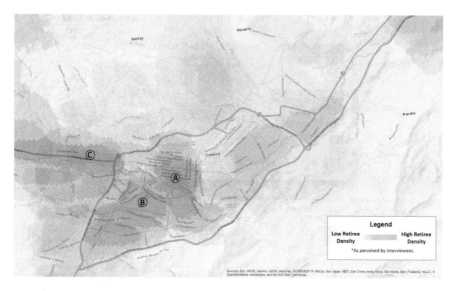

**Fig. 3.2** Density map of Cuenca, Ecuador, showing the areas with higher and lower concentrations of immigrant retirees from North America. A = the historic city center; B = the newer area of the city adjacent to the city center; C = "Gringolandia" – an area of high-rise apartments and shops that are know to cater to immigrant retirees

the map is the city center – a dense zone of commercial buildings and low residences, located on an elevated plain north of the river Tomebamba. Living in this area is not particularly popular among Cuencanos, in large measure because it is noisy and parking is practically impossible to find, but has become popular with retirees. Neighborhood B is the "new" hub of the city, a large, mixed use area of small homes, high rise apartments, businesses, and restaurants that is particularly popular with immigrant retirees. Located on a lower, flat area south of the Tomebamba river and immediately across from the city center, it is a vibrant mixture of businesses, small homes, apartments, and high rises. Neighborhood C is a small strip of high rise apartments along the Avenida Ordóñez Lasso, many of which overlook the picturesque, park-like Tomebamba river, a short walk or bus ride west of the city center. This small region has the highest concentration of immigrant retirees, and so – though the majority of its residents are Ecuadorian – it is universally known as "Gringolandia."

Multi-story, secured-access high-cost apartment buildings are a feature of the real estate landscape in Cuenca, and a common residence choice for immigrant retirees. Like scattered trees towering over bushes in an arid landscapes, the high-rise apartments are easy to spot (Fig. 3.3). This type of dwelling had begun to be popular in Cuenca before the immigrant retirees began to flow in great numbers, and even now the vast majority of apartment units are occupied by Cuencanos. A typical high rise in Gringolandia is pictured in Fig. 3.4.

Figure 3.5 locates the residences in and around Cuenca of the 300 expat retirees who responded to our survey (dots on the map), and also the residences of the 38

**Fig. 3.3** In this photograph of the Cuenca Skyline, it is easy to spot the newer, fancier high-rise apartment complexes that tend to attract immigrant retirees, as well as well-to-do Cuencans and natives who have returned from Europe of North America

native Ecuadorians we interviewed, many of whom would be considered upper middle class or above (crosses on the map). The expat retirees are largely concentrated in and around the three central areas identified earlier (Fig. 3.2), whereas two-thirds of the native Ecuadorians were residing in areas peripheral to these central zones.

Not all immigrant retirees live in the areas of concentration, however. A scattering live northeast of town, in a picturesque valley called Challuabamba, 15 minutes by freeway from the city proper. Scatterings of retirees are also found in other areas – both neighborhoods within the city and outlying towns. Most, however, live in the three central zones.

Due to the fact that immigrant retirees are approximately 2% of the population of Cuenca, and the majority rent their residences (Fig. 3.1), only a small minority of dwelling units in Cuenca are owned by the expat retirees. "Extranjeros," according to one realtor we interviewed, own about 10% of the city center and are responsible for almost all major apartment or home renovations in that zone, but city-wide they own only 4% of units.

**Fig. 3.4** Apartment building in Gringolandia (Cuenca, Ecuador)

San Miguel de Allende. Figure 3.6 is a density cloud map of San Miguel de Allende, indicating where in the city immigrant retirees tend to live. Compared to Cuenca, the city is smaller and more concentric around its central plaza, and retiree concentrations tend to radiate in all directions from the city center, gradually expanding the "gringo" zones as the retiree population continues to grow. Neighborhood A is the Centro, the old colonial center of the city, with the highest real estate prices, a still-desirable area in which homes and apartments have to some extent given way to hotels, restaurants, and short-term rental apartments. Neighborhoods B (San Antonio) and C (Guadiana), both close to the Centro, are perhaps the areas of town with the densest concentration of immigrant retirees. Neighborhoods D (Guadalupe) and E (Independencia) are areas that until recently were largely occupied by local families, but that are been rapidly gentrifying.

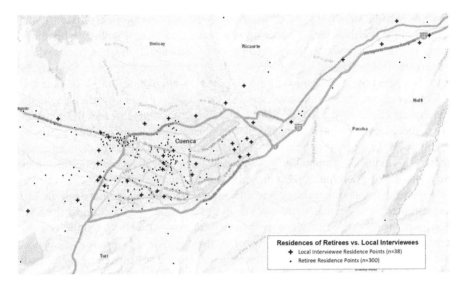

**Fig. 3.5** Map of Cuenca demonstrating the residences of immigrant retirees who responded to our online survey (dots) and local residents we interviewed who provided their residence location (crosses). Overall considerable co-location is evident, though there is a tendency for local residents we interviewed to live more frequently in the periphery and for immigrant retirees to live in the three central areas noted in Fig. 3.2

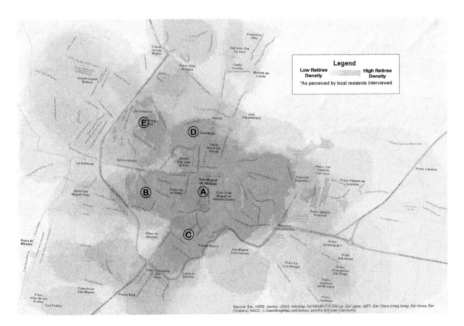

**Fig. 3.6** Density map of San Miguel de Allende, Mexico, showing areas with higher and lower concentrations of immigrant retirees from North America. The neighborhoods with letters A–E have the highest concentrations of immigrant retirees

Further outside are several other neighborhoods that, because of proximity, the quality of the housing stock, or development, have also become popular with immigrant retirees. There are no true gringo ghettos – virtually all neighborhoods where Americans live are mixed.

A few gated communities exist (Fig. 3.7) and tend to include both Mexicans and Americans. The one possible exception is Los Labradores, an upscale gated community approximately 5 miles from San Miguel de Allende, near the town of Atotanilco. Developed and managed by a company from Toronto, Canada, it is in many ways like a continuing care retirement community in the states, with approximately 100 homes, a clubhouse, and activities for independent older persons, plus an assisted living community that provides care all the way to death.

Figure 3.8 locates the residences in and around San Miguel de Allende of the 210 expat retirees who responded to our survey (dots on the map), and also the residences of the 43 native Mexicans we interviewed, many of whom would be considered upper middle class or above (crosses on the map). As was the case in Cuenca, the expat retirees are largely concentrated in and around certain central areas, with the residences of the majority of native Mexicans located in a circular penumbra of neighborhoods more distant from these central zones. As in Ecuador, a scattering of retirees can be found on and beyond the outskirts of the city, often in new communities developed for gringos and more well-to-do Mexicans. However, the vast majority of immigrant retirees live in the city proper.

**Fig. 3.7** Entrance to a gated commuity in San Miguel de Allende

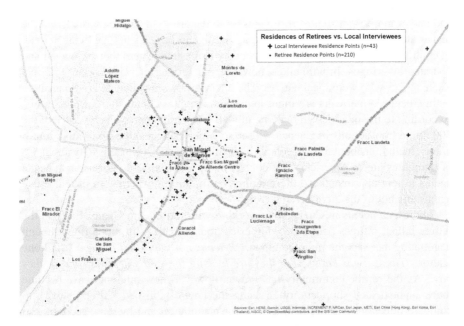

**Fig. 3.8** Map of San Miguel de Allende demonstrating residences of immigrant retirees who responded to our online survey (dots) and local residents we interviewed who provided their residence location (crosses), showing a tendency of the immigrant retirees to concentrate near the city center and the local residents we interviewed to be located more distally

Indeed, the high cost of real estate and a relative absence of vacant land in the city proper has led to marked growth of communities on the outskirts, mostly for natives who have either sold their homes in the city or were priced out of the market. Many of these communities lack the services that are provided in the city proper. For example, a couple of these fringe neighborhoods lacked running water for years; instead, the town arranged to have a truck bring water to the neighborhood on a regular basis. Other of the satellite communities are known for crime and gang violence, as police resources are reported to be focused on more central neighborhoods. Indeed, as will be discussed later, the city is considering building an entire community outside of town, to make low cost housing more available for natives who work in town but can't afford to live there.

## 3.3 What Local Real Estate Agents Say About Retirees

We formally interviewed six real estate agents in each city. These were convenience samples, chosen by a combination of networking, looking on the internet, and finding individuals who were willing to be formally interviewed. All had to be natives of the country where we were doing the research and be working for a company that had one or more English advertisements on the internet. In Cuenca virtually all the real estate agents are Ecuadorian, though some agents have spent time living abroad. In contrast, San Miguel de Allende has many real estate offices that are owned and operated by gringos. In both cities, however, our formal interviews were with real estate agents who were local natives. The interviews were conducted in Spanish.

In Cuenca, Americans are more important as rental clients than as purchasers. According to the interviewed realtors, in a given month, approximately 20% of the 300 rental transactions in Cuenca will be to Americans or other non-native immigrants, including about 50% of rentals in the historic center; whereas only about 5% of purchases are to Americans. Of course, many Cuencanos do not use real estate agents to rent apartments, so these figures represent the experiences of real estate agents and not of the city overall.

Real estate agents in Cuenca like working with retired Americans. Early in the retiree influx, retirees paid the asking price; today, that is still the norm, although negotiating has become more common. *"Price is not important to them,"* said one realtor. *"The gringos know what they are looking for, and when they find it, they pay."* As the retiree community has grown, however, new migrants have become more sophisticated shoppers, often using the now-established retiree network to compare prices, providing them a basis for negotiation and thereby, according to realtors, becoming more like Ecuadorians. Still, the opinion among non-real estate agent interviewees prevails that American retirees are desired customers and that real estate agents prefer to sell to immigrant retirees (Table 3.1). As one realtor justi-

**Table 3.1** Responses of Local Resident Interviewees to Quantitative Interview Questions about Impact of Immigrant Retirees on Real Estate. Scores range from 0 (complete disagreement) to 4 (complete agreement)

| | Respondents | | | | | |
|---|---|---|---|---|---|---|
| | All respondents[b] | | | Real estate agents only, by city | | |
| Statement [English Translation] | Real estate agents | All others | P–value[a] | Cuenca | San Miguel de Allende | p-value [a] |
| Real estate prices have increased because of the demand of retired immigrants | 1.00 | 1.20 | .555 | .667 | 1.33 | .188 |
| Real estate agents prefer to sell or rent to immigrant retirees | 2.50 | 3.37 | **.007** | 2.00 | 3.00 | .289 |
| Foreigners are causing local residents to move from the center of the city to more outside areas of the city | 2.25 | 2.39 | .736 | 1.17 | 3.33 | **.015** |
| Parts of our city are an American Colony | 2.42 | 2.47 | .885 | 3.33 | 1.50 | **.012** |

[a]Determined using t-test for difference in means
[b]Total – 78 respondents, of whom 17 were government workers, 12 were service providers, 13 were health care personnel, 12 were real estate agents, and 24 were tienda owners, equally divided between the two cities

fied, *"An immigrant retiree couple will take care of the apartment. They will keep it clean. They won't mark the walls..."*

Retired Americans who move to Cuenca generally want to live near the historic center of town. This provides them with ready access to restaurants, coffee shops and cultural sites such as the symphony (which is free of charge). Other preferences of American property buyers, according to the realtors we interviewed, include: apartments with security; apartments that overlook a river or have a nice view of the mountains; new construction; large bedrooms; a kitchen that is large enough to serve as a social area; balconies or terraces; lots of light; location near a supermarket; luxurious materials such as granite or marble; and floating floor systems of such materials as bamboo.

In addition, several realtors we interviewed in Cuenca talked about the importance to many retired Americans of their pet dog or cat, such that access to a park was a major element in their choice of a place to live. One realtor, for example, talked about having a pet as a hallmark of the immigrant retirees: *"They come accompanied by a little pet, which is a trademark of the retired American. For them, a little dog or cat is part of the family."*

In San Miguel de Allende, where the American presence is so much more prominent, the real estate market is divided in two. Realtors we interviewed spoke of half of the market – including virtually all of the higher priced properties – being conducted in US dollars. The other half is conducted in pesos. American retirees, international investors attracted by the tourist market, and some wealthy Mexicans pay in dollars; the vast majority of locals pay in pesos. Americans often pay cash; virtually all Mexican buyers purchase on credit. So large is the dollar real estate market that most real estate agencies have at least one American on staff. About a quarter of the city's approximately 170 real estate agents are American, and a number of companies are American owned and operated. One Mexican realtor we interviewed mentioned with disapproval that some expat real estate agencies join the local real estate association but refuse to share their sales with their Mexican colleagues. She felt that this type of behavior was contrary to the collegial spirit of the native realtors.

In spite of the many ways in which real estate in San Miguel differs from that in Cuenca, the preferences of immigrant retirees, as explained by the realtors, were remarkably similar. Living near the center of town is a top priority and the most commonly reported characteristic of immigrant retirees, according to the realtors in San Miguel de Allende we interviewed. However, with the increasing influx of tourists, especially on weekends, many retirees find the city center less attractive and are choosing to live in quieter neighborhoods. An additional characteristic that was frequently mentioned was desire of retirees to be in a mixed neighborhood – this being a given in Cuenca and indeed largely a given in San Miguel de Allende as well, where integration of Mexican and expat residences is largely the norm as well. Other common preferences of retirees that our interviewees mentioned are: walking distance to restaurants and activities, a bedroom on the ground floor, access to the outdoors by way of a patio or terrace, good water pressure, heat for winter, and proximity to a park where they can take their pet.

Whether a transaction is in Cuenca or San Miguel de Allende, selling to an American takes more time and effort from the realtor than selling to a native. Between their lack of language skills and their unfamiliarity with the local bureaucracy, Americans need a lot of hand holding to negotiate the administrative hoops associated with purchasing a home. Many get some help from a "facilitator" – a local who, for a fee, helps the retiree carry out various types of business. Typically, however, the realtor gets involved. One realtor in Cuenca put it this way:

> *"When you sell a house to an Ecuadorian, he can continue the process on his own. But a gringo, no. You have to hold their hand. Sign here; sign here; sign here; sign here. From one building to another, you have to guide them through the entire bureaucracy."*

## 3.4 Housing Prices

To obtain general and comparative data on real estate pricing in Cuenca and San Miguel de Allende, we identified all properties listed as for sale or for rent on selected internet sites during October 8–12, 2018, with the goal of gathering data on at least 200 consecutive listings in each location. Sites were identified by Google searches for "inmobiliaria [city/country]" and "bienes raices [city/country]". Selection of sites for data collection from those identified by our internet search was based on three criteria: size (preferring the site with the largest number of listings, as such sites involve listings from multiple companies and agents); listing in Spanish (to survey a more general sample of units); and provision of an exact or general geo-location for each listing (to provide us with the ability to do sub-analyses by region within a city). By requiring that sites provide geo-location of listed properties, the list of eligible websites was considerably reduced, as this service was not the norm for most websites.

We obtained information on residential property for sale in Cuenca using plusvalia.com, the largest real estate portal in Ecuador, from which we were able to obtain listing information on 200 consecutive properties. Similarly, information on residential property in San Miguel de Allende was obtained from a single multiple listing site, propiedades.com, from which we obtained listings on 233 properties. Rental property listings in both cities were obtained from multiple sites. In Cuenca we used plusvalia.com, olx.com.ec, and mercadolibre.com.ec; a total of 58 listings were obtained from these sites. In San Miguel de Allende we obtained data from the appropriate subsections of propiedades.com, icasas.mex, and ocanto.com, from which we obtained 39 rental listings. Commercial property listings were, of course, not included in our samples. Also, due to limitations in the availability of information, we were unable to differentiate condominium apartments from freestanding dwellings.

The data we gathered during that survey are displayed in Table 3.2. It shows that in both cities the variation by neighborhood is relatively modest, a verification of the

**Table 3.2** Price listings of real estate listed "for sale" on selected internet sites in the two study cities[a]

**San Miguel de Allende**

|  | City-wide (n = 233) | Neighborhoods popular with U.S. retirees | | | | | Other neighborhoods (n = 112) |
|---|---|---|---|---|---|---|---|
|  |  | City center (n = 65) | San Antonio (n = 23) | Guadiana (n = 9) | Guadalupe (n = 10) | Independencia (n = 14) |  |
| Median price | $238,770 | $238,770 | $282,810 | $254,688 | $242,500 | $242,038 | $214,308 |
| Median size (m²) | 219 | 250 | 154 | 231 | 200 | 228 | 209 |

**Cuenca**

|  | City-wide (N = 199) | Neighborhoods popular with U.S. retirees | | | Other neighborhood (n = 151) |
|---|---|---|---|---|---|
|  |  | City center (n = 15) | New city (n = 18) | "Gringolandia" (n = 15) |  |
| Median price | $165,000 | $295,000 | $245,000 | $165,000 | $150,000 |
| Median size (m²) | 192 | 210 | 184 | 220 | 185 |

[a]Data gathered during a single week in October 2018

fact that segregation by wealth has not traditionally been a feature of Latin American cities, and this mixing of well-to-do and more modest homes continues in large measure within these cities. The data also make clear that the average price in San Miguel de Allende is nearly 50% higher than in Cuenca. In both cities, of course, the cost of housing varies markedly within and across neighborhood and whether the property is new or recently renovated, or old and in need of repair.

## 3.5  Increases in Real Estate Prices and the Cost of Living – Are Retirees Responsible?

Housing price inflation was a common theme of interviews with local real estate agents in both Cuenca and San Miguel de Allende. Five of our six interviewees in Cuenca specifically talked about prices having increased, especially for the more expensive properties. In San Miguel de Allende, rising prices are a given, and the main topic of conversation around price was how much the price inflation has led to a flight of locals from the city center. What was unclear over the course of interviews was whether the benefit of the additional income in these areas due to increased tourism and retirement outweighs the effects of gentrification, and, if so, to what extent retiree immigration is responsible.

In Cuenca, real estate prices have risen throughout the city over the past decade. However, most American retirees we met rejected the idea that they've had significant impact on the cost of living in general or of housing in particular. "We are only one percent of the population; we're too few to be responsible," was the typical response in-person; however, many retirees within our online survey pointed to real estate costs as a primary concern.

Indeed, data from the government of Ecuador's annual cost of living survey indicate that Cuenca's rental market was the most expensive in the country in 2009, before the American retiree boom. Rental prices have largely risen at the same rate in Cuenca as in other cities during the decade that followed, except in 2014–2016 when prices did increase faster than in most other cities (Fig. 3.9).

Are American retirees responsible for price increases in Cuenca? Most likely they contribute some, particularly by increasing demand for high end apartment units and for housing near the city center. On the other hand, it seems clear that Ecuadorians living abroad or returning from abroad with investment cash are a bigger factor than the American retirees. One real estate agent put it this way: *"About two and a half million Ecuadorians live outside the country. They send money back with the sole objective of buying real estate and automobiles."*

In San Miguel de Allende, real estate prices have been going up for decades, and in recent years this has accelerated. Indeed, prices in the central neighborhoods have risen so steeply that they are said to be "comparable to Texas or California." One realtor we interviewed in the summer of 2018 noted that the market was so hot that fewer houses were available than buyers. *"Right now, our inventory of homes for sale is a bit low. A lot of people want to invest, but there are few options."* As a result, no one nowadays chooses the city because it is an affordable place to live. Indeed, in spite of retaining its colonial feel, the city has become more like Cabo san Lucas or Cancun than a typical Mexican town.

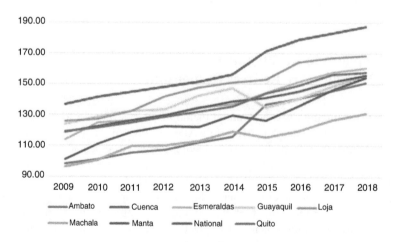

**Fig. 3.9** Monthly rental cost for a family of four in Cuenca in U.S. dollars, May 2009 – May, 2018. Source: Instituto Nacional de Estadística y Censos, Ecuador

Are immigrant retirees responsible for the boom in real estate prices in San Miguel de Allende? Certainly, at something like 10% of the population (Fernández 2011), it's indeed likely that they contribute. Our interviews indicated, however, that at least three other factors may be even more important. One is the growth of Airbnb, VRBO, and other sites that make it profitable for someone to buy a property, live in it some of the time, and rent it out when tourist or "snowbird" demand is high. Indeed, four of our six real estate interviewees specifically implicated the popularity of this part-time rental phenomenon as a major cause of price inflation. Given the city's current popularity as a tourist destination – 2018 being the second year in a row that the popular English language magazine *Travel and Leisure* listed the city as the number one tourist city in the world (Terzian 2018) – real estate investment by non-retirees clearly has a market.

The other non-retiree factor driving up housing prices in San Miguel de Allende is Mexicans from other parts of the country who look to the city for second homes, retirement homes, or investment. Three of the six real estate agents we interviewed in San Miguel de Allende felt that investors from other Mexican cities like Querétaro and Mexico City were heating up the real estate market as much as, and possibly more than, the Americans.

Whatever the reason, the increase in real estate costs is causing hardship for locals. According to one realtor in Cuenca, this means that occasionally two families need to move in together in order to afford a home, whereas for the immigrant retirees paying the rent does not pose a problem:

> *"At times two families now have to live together in a house, in order to pay the rent, whereas for an immigrant retiree paying the rent is easy."*

The impact of price inflation was noted even more strongly in San Miguel de Allende, where it is widely acknowledged that service workers have been completely priced out of housing in the city. *"The cost of everything has gone up, but not salaries,"* the housekeeping supervisor in a rental complex told me.

## 3.6 Governmental Role in Promoting International Development as a Factor in the Real Estate Boom in Both Cities

In both cities the local governments have embraced a growth and development process that is welcoming to outsiders, including retirees. Designation of the city's historic center as a UNESCO World Heritage site, the culmination of an arduous application process which both cities have achieved, represents one example of this active governmental involvement in international promotion (United Nations Educational, Scientific, and Cultural Organization (UNESCO) 2017). But UNESCO designation is not enough – all told, 51 cities in Latin America have UNESCO World Heritage designation on the basis of having historic colonial architecture and a preservation plan for the designated area (United Nations Educational, Scientific, and Cultural Organization (UNESCO) n.d.); – yet only a few have attracted migrants to the extent of Cuenca or San Miguel de Allende.

A regional employee of an agency dedicated to recording statistical and geographical data noted that several nearby historic colonial cities in Mexico have UNESCO World Heritage status and should, therefore, be potentially able to attract retirees. Examples are Zacatecas, Guanajuato, and Morelia. He then went on to talk about why SMA is unique, and what it might take to bring the other cities to have similar success in attracting immigrants from the north. Some of his reasons include: cleanliness of the city (no garbage, freshly painted houses), areas of safety (away from crime-related routes, as well as adequate policing and security measures), investment of money into activities that appeal to retirees (like the arts), economically healthy outlying areas, as well as aggressive promotion as a tourism and destination site.

Indeed, San Miguel de Allende has a long history of promotion as a tourist and retiree destination, beginning in the early 1950s, when former Guanajuato governor Enrique Fernández Martínez founded the Instituto Allende to encourage foreigners to come to study art. Cuenca began this process in earnest only a decade or so ago, during a tourism promotion campaign undertaken by the municipality during 2006–2007 (Pallares and Rollins-Castillo 2019). This, in conjunction with early retiree immigration and outside media attention, launched the municipality's interest and commitment to continuing to promote the city as a destination for retirees. In recent years, governmental promotion efforts have continued and if anything accelerated. Recently two grandiose projects illustrate the lengths to which local governments will go to promote tourism and retiree immigration:

- construction of a light railway (tram) from the airport through central Cuenca, through Gringolandia, and extending to the southeastern reaches of the city), a project that continues to move forward despite delays and cost overruns (CuencaHighLife 2012); and
- a recent plan to develop a large, low-cost satellite city outside of San Miguel de Allende, to house locals who work in the city but cannot afford to live there (Aguado n.d.; López 2019; News San Miguel n.d.).

## 3.7 Displacement of Local Residents from the City Center to Outlying Neighborhoods

Displacement of local residents was a common theme of real estate agents in San Miguel de Allende but not in Cuenca. In our in-depth interviews with six real estate agents in each city, four in San Miguel de Allende talked about displacement of residents, often at great length. In contrast, no real estate agents in Cuenca mentioned displacement of locals within the city due to American retirees. These findings were mirrored by the results of the objective questions included in the interviews (Table 3.1), in that realtors in San Miguel de Allende were significantly (p = 0.015) more likely to assert that foreigners are causing local residents to move from the center of the city to more outside areas of the city.

Not all of the displacement of locals in San Miguel de Allende is due solely to retirees, however. The city's growth as a tourist destination has been an important factor, rendering the city's center more noisy and commercial than in the past:

> *"The retirees and the tourists have made the city center more commercial than residential. It's rather a tourist destination, with its restaurants, hotels, bars, and shops. But at the same time there are fewer houses, fewer places to live."*

Additionally, it is a natural phenomenon, according to one interviewee, that over time an old family homestead will become occupied by a widow or widower, and when that person dies, it makes sense to the family to sell the property, providing cash to help multiple family members economically, rather than continuing to keep up with the expenses of maintenance. As explained by a tienda owner in an outlying neighborhood of San Miguel de Allende, *"the houses in the center of town went up in price and in appreciated value; so many people who had lived for years in the city center retired, sold their homes, and stopped living in the city center, to improve the economy of their families."* Another interviewee similarly spoke of his family's experience of selling a home that had been theirs for generations:

> *"We rented part of our house, but the income was not sufficient for us to pay to remodel the house or to build up savings. So, we decided to sell the property, and that allowed my brother to build his house in Guanajuato and for me to build the house where I now live with my family. I don't think of this as a problem; instead it helped me."*

However, not all interviewees felt this way. Some expressed resentment at the difficulty experienced in trying to keep their ancestral homes in the family. For example, one local resident in San Miguel spoke of the difficulty the local population had in overcoming bureaucratic regulations related to home renovation compared to what she felt was a much more facilitative approach of the municipality turned towards immigrant retirees. This impression of local bureaucracy's lack of accountability when working with foreign entities has been reflected in the academic literature as well – in San Miguel de Allende (Pacheco and Vallejo 2016), as well as in other retirement locations like Granada, Nicaragua (Foulds 2014).

While acknowledging that displacement of local residents was occurring, real estate agents in San Miguel de Allende had mixed opinions about whether this trend was good or bad. Several talked about how the increased noise and traffic had made the city center less desirable. Others emphasized that these changes were accompanied by an economic boom that provided many work opportunities.

In summary, opinions of local residents we interviewed revealed a complex, nuanced attitude, which acknowledged both the positive gains that families had experienced, as well as the losses.


## 3.8 Gentrification and Price Inflation – Are Migrant Retirees Responsible?

In our two study cities of Cuenca, Ecuador and San Miguel de Allende, Mexico, gentrification is clearly occurring. "Desirable" areas in or near the center of the city are becoming more expensive; historic homes are being remodeled or rebuilt; and businesses catering to international, often well-to-do, tastes – such as supermarkets, organic groceries, and art stores – are proliferating in certain areas. Concomitant with these changes is relocation of many natives to less expensive, ungentrified, peripheral residential areas.

There are also pockets of locals holdouts against gentrification, even adjacent to neighborhoods popular with retirees. One such holdout neighborhood is up a hill from the city center in San Miguel de Allende, where none of the generations of family have sold their property to foreigners or wealthy Mexicans. Residents of this small neighborhood, the Valle de Maiz (Valley of Corn) are described by real estate agents as "stubborn" and "traditionalists." However, their status appears at risk, in part because a local festival has become trendy, and in part because roadway improvements near the area are making it more of interest to realtors and buyers. According to a May, 2018 notice in the San Miguel Times, "being one of the few areas in town where one could go 'into the woods' if only for moment, I fear these tree-laden dog walking adventures are coming to a close." (Toone 2018)

Gentrification is often criticized because of its effect on native communities. One argument is that people displaced by gentrification are forced to relocate to less desirable neighborhoods, where there may be fewer municipal services. We heard this much more in San Miguel de Allende than in Cuenca, due no doubt to the more limited impact of both immigrant retirees and gentrification in the latter city. We noted in San Miguel that some of the peripheral communities had markedly fewer services – one, for example, did not have municipal water; instead, for years water was brought by truck to central areas of the community. In others, we heard stories of high levels of gang activity and youth delinquency. These results from interviews regarding a lack of services in some areas have been reflected in census data as well. CONEVAL (Consejo Nacional de Evaluacion de la Politica de Desarrollo Social), a public organzation that collected census data in 2010, found that 63.7% of residents of San Miguel de Allende lived in poverty, and a significant number of households lacked access to basic services and food; particularly within the outskirts of the city center (Pacheco and Vallejo 2016).

Another complaint about gentrification is that it accentuates wealth disparities by driving poorer residents out of an area, leading to de facto segregation by wealth.

With increasing income disparities globally, this has become an issue worldwide. Among our study cities, it was most evident in San Miguel de Allende. The effects of this wealth segregation can be seen clearly in the current planning of a satellite community approximately 6 km from San Miguel de Allende that is intended as affordable housing for natives who work in the service sector (Aguado n.d.; López 2019; News San Miguel n.d.).

These issues can create tensions in the community. The most telling example we saw was in San Miguel de Allende. A golf course and residential community were developed several years ago on the northern outskirts of town. On the far side of the golf course sits a relatively poor, almost entirely native community called San Luis Rey, that is known to have gangs and where wall graffiti is prominent. Developers wanted to build high-end homes on the fringe of San Luis Rey, overlooking the golf course. So they have built a large, long wall to separate the new homes from the local community. According to one local resident we interviewed, in order to assuage the community, the developers had paid local nonprofits and artists, as well as, reportedly, local gangs, to create graffiti murals as a town beautification activity (Artesanto 2019; CASA n.d.).

Whether and to what extent gentrification is due to retirees is debatable. Our interviews and survey did make it clear that gentrification in the old historic area of both cities has been to a significant degree brought about by foreigners. However, other than in the small historic center, and even to an extent in that area, others have had a major role in gentrification. In San Miguel de Allende, well-to-do Mexicans have clearly had a significant role, coming for the same reasons as the retired Americans and often wanting similar amenities, such supermarkets and international restaurants. In Cuenca that role is filled by returning Ecuadorians, who have come back with saved money to buy homes and start businesses, and who too have acquired a taste for international cuisine.

In San Miguel de Allende especially, many interviewees agreed that displacement of locals has been a consequence of the retiree migration and price inflation. Many did not, however, view this trend as negative. They compared their city to nearby cities that were struggling economically – a situation in which everyone suffers. In that context, they explained, having economic growth and a healthy job market was worth the trade-off, especially since people who move to the outskirts can still work in the city with its higher wages and greater opportunities.

## 3.9  Discussion and Conclusions

Over the course of our data collection, it became apparent that local residents generally believe that foreign retirees have impacted each city's real estate landscape. Interviewees identified a number of ways in which retirees have made their mark on the housing within the cities – including altering real estate agents' renting preferences, creating increased demand for housing in historical neighborhoods, contributing to rising housing costs, and in doing so, contributing to a gentrification and displacement process within the cities.

In both cities, retirees tend to congregate within a few areas and neighborhoods around the historic city center. These "settlements", with higher numbers of foreign residents than average, have been replicated in other retirement destinations, including Swedish retirees in Spain, where entire complexes and neighborhoods can be dominated by immigrant retirees (Gustafson 2016).

The gentrification evident in both cities includes not just the reclamation of historic homes as housing for foreign residents, but changes in neighborhoods near those areas as well. In Cuenca and in San Miguel de Allende, the city centers have new bars and cafes catering to tourists and lifestyle migrants that advertise in English, employ English speaking staff (which they can charge higher prices for), and host events for foreigners. As prices rise in these establishments, the area becomes less affordable for local residents, instead catering to those who have the purchasing power to modify the culture of the historical center to their interests and needs (van Noorloos and Steel 2016; Janoschka et al. 2014; Steel and Klaufus 2010; Pacheco and Vallejo 2016; Navarrete 2018; Hayes 2015). Respondents within our research gave nuanced opinions regarding their thoughts on this type of gentrification. Some residents spoke to the benefits they were able to reap by selling their homes, or the benefits they believe this tourism and change bring to the general economy; others, more commonly within San Miguel de Allende, spoke to the frustration of the increased housing prices placing them out of homes and neighborhoods that had been within their families' for generations. The benefit that comes to the local population from selling their property is debated as well. Some literature points to the fact that a local resident may sell their home to a foreign buyer for an artificially low price, which in the end benefits the foreign buyer far more, when the value continues to increase (Navarrete 2018).

Additionally, the majority of interviewees felt that real estate agents within their respective cities preferred to rent or sell to immigrant retirees. As was discussed in Chap. 2, this was an impression held by both immigrant retirees and local real estate agents in both cities. Prior research in Cuenca found similar themes; a former real estate agent interviewed felt that landlords preferred renting to foreigners due to the idea that retirees will provide punctual payments, will take better care of the property, and will be willing to pay higher rates (Bustamante et al. 2012).

The extent of the impact of retirement migration on the real estate market within Cuenca and San Miguel de Allende differs for a number of reasons. Cuenca is a larger city, with proportionally fewer retirees and a modest tourist presence in general, compared to San Miguel de Allende, a smaller city with high numbers of retirees, second-home owners, and tourists. So, although gentrification is clear within both cities, the magnitude of gentrification and displacement in Cuenca is modest compared to San Miguel de Allende, where the topic was a much greater concern among local residents we interviewed.

An additional difference between the two cities is that retirees in each displayed markedly different housing preferences. Retirees in Cuenca were more likely to rent an apartment than to buy a home; the opposite was the norm in San Miguel de

Allende. A number of factors may cause this difference: the different financial situations of the retirees, the available housing market in each city, and the investment potential of real estate. Additionally, research has found that some migrants find renting gave them more freedom, eliminating the pressures and responsibilities of home ownership (Gustafson 2016; Bustamante et al. 2012); so the differential preferences in the two cities may reflect not just income differences but also variation in interest in this aspect of putting down roots.

This is not to say that local residents consider retirees to be solely responsible for the real estate changes that have been occurring in these cities, an observation that has been made by others as well (Schafran and Monkkonen 2011; Hayes 2018). Within Cuenca, discussions of rising costs included mentions of Ecuadorians living abroad and investing in the city and returning with resources to purchase or rent high-end homes. In San Miguel de Allende, considerations included a boom in tourism-related conversions to Airbnb, VRBO, and other rental options, as well as persons buying second homes or investment homes. Put together, these forces reflect an increasingly cosmopolitan and "transnational" aspect to these cities, which in itself is a reflection of globalization as a social and economic force (Hayes 2015).

In summary, changes in the real estate markets in Cuenca and San Miguel de Allende are complex, with the presence of retirees from the United States, Canada, and to a lesser extent western Europe being just one of many factors. From our interviews and data collection, it is evident that both of these cities are undergoing similar changes in their real estate landscapes – including increased demands for more luxury housing, increased real estate prices, and changing use of buildings in the historic city centers. However, to what extent these changes are due primarily to foreign retirees remains unclear.

## References

Aguado, J. (n.d.). Lomas de San Miguel could be the biggest and most populated neighborhood in the city. *Atención San Miguel.* http://www.atencionsanmiguel.org/2017/12/01/lomas-de-san-miguel-could-be-the-biggest-and-most-populated-neighborhood-in-the-city/. Accessed 1 Dec 2017.

Álvarez, M. G., Guerrero, P. O., & Herrera, L. P. (2017). Estudio sobre los impactos socioeconómicos en Cuenca de la migración residencial de norteamericanos y europeos: Aportes para una convivencia armónica local. In *Informe Final.* Cuenca, Ecuador: Avance Consultora.

Artesanto. (2019, January 22). La Simbología detrás de los murals de la colonia San Luis Rey. *Artesanto: San Miguel de Allende.* http://artesanto.mx/la-simbologia-detras-de-los-murales-de-la-colonia-san-luis-rey/. Accessed 15 July 2019.

Bastos, S. (2014). Territorial dispossession and indigenous rearticulation in the Chapala Lakeshore. In M. Janoschka & H. Haas (Eds.), *Contested spatialities, lifestyle migration and residential tourism* (pp. 63–75). Abingdon: Routledge.

Bayer, J. (2018). A guide: Relocating to Ecuador. *Abundant Living, Ecuador.* https://docs.wix-static.com/ugd/dfaf53_3f2044028c9d48dca75e2a7a099497c1.pdf. Accessed 26 July 2019.

Bernier, E. T. (2003). El turismo residenciado y sus efectos en los destinos turísticos. *Estudios Turísticos, 155/156,* 45–70.

Bustamante, A. V., Laugesen, M., Caban, M., & Rosenau, P. (2012). United States-Mexico cross-border health insurance initiatives: *Salud Migrante and Medicare in Mexico. Revista Panamerica de Salud Pública. SciFLO Public Health, 31*(1), 74–80. https://www.scielosp.org/article/rpsp/2012.v31n1/74-80/en/. Accessed 26 July 2019.

CASA. (n.d.). Murales de San Luis Rey. *CASA.* https://casa.org.mx/murales/. Accessed 15 July 2019.

Casado-Diaz, M. A. (1999). Socio-demographic impacts of residential tourism: A case study of Torrevieja, Spain. *International Journal of Tourism Research, 1*(4), 223–237.

Chaca, S. (2018). Buy property in Ecuador: Here's what you need to know. *Cuenca High Life.* https://cuencahighlife.com/buying-property-in-ecuador-heres-what-you-need-to-know/. Accessed 29 Nov 2018.

CuencaHighLife. (2012, October 3). Cuenca's new light rail system, Tranvía de los Cuatro Rios, is scheduled to be operational in 2014. *Cuenca High Life.* https://cuencahighlife.com/cuencas-new-light-rail-system-tranvia-de-los-cuatro-rios-is-scheduled-to-be-operational-in-2014/. Accessed 26 July 2019.

David, I., Eimermann, M., & Akerlund, U. (2015). An exploration of lifestyle mobility industry. In K. Torkington, I. David, & J. Sardinha (Eds.), *Practicing the good life: Lifestyle migration in practices* (pp. 138–116). Newcastle upon Tyne: Cambridge Scholars Publishing.

Fernández, I. G. (2011). The right to the city as a conceptual framework to study the impact of north-south migration. *Recreation and Society in Africa, Asia and Latin America., 2*(1), 3–33.

Foulds A. (2014). Buying a colonial dream: The role of lifestyle migrants in the gentrification of the historic center of Granada, Nicaragua. *Theses and Dissertation – Geography, 18.* https://uknowledge.uky.edu/geography_etds/18/. Lexington: University of Kentucky.

Global Property Guide. (2017). Total transaction costs range from low to high in Mexico. *Global Property Guide.* https://www.globalpropertyguide.com/Latin-America/Mexico/Buying-Guide. Accessed 29 Nov 2018.

Gregor, A. (2014). House hunting in… Ecuador. *New York Times.* https://www.nytimes.com/2014/10/16/realestate/real-estate-in-ecuador.html. Accessed 21 Feb 2018.

Gustafson, P. (2016). Our home in Spain: Residential strategies in international retirement migration. In M. Benson & K. O'Reilly (Eds.), *Lifestyle migration: Expectations, aspirations and experiences* (pp. 69–86). New York City/Oxon: Routledge/Taylor & Francis Group.

Haines, B. (2018a). Ecuador real estate guide. *Gringos Abroad.* https://gringosabroad.com/ecuador-real-estate/. Accessed 29 Nov 2018.

Haines, B. (2018b). 7 tips for expats buying property in Ecuador. *Gringos Abroad.* https://gringosabroad.com/buying-property-in-ecuador/. Accessed 29 Nov 2018.

Haines, B. (2018c). How to rent or buy property in Cuenca Ecuador. *Gringos Abroad.* https://gringosabroad.com/how-to-rent-or-buy-housing-in-cuenca/. Accessed 29 Nov 2018.

Hayes, M. (2015). Moving south: The economic motives and structural context of North America's emigrants in Cuenca, Ecuador. *Mobilities, 10*(2), 267–284.

Hayes, M. (2018). *Gringolandia: Lifestyle migration under late capitalism.* Minneapolis/London: University of Minnesota Press.

Janoschka, M., Sequera, J., & Salinas, L. (2014). Gentrification in Spain & Latin America – A critical dialogue. *International Journal of Urban and Regional Research, 38*(4), 1234–1265.

Lizárraga-Morales, O. (2008). Immigration and transnational practices of US retirees in Mexico. A case study in Mazatlán, Sinaloa and Cabo San Lucas, Baja California Sur. *Migración y Desarrollo, 11,* 93–110.

Lizárraga-Morales, O. (2010). The US citizens' retirement migration to Los Cabos, Mexico. Profile and social effects. *Recreation and Society in Africa, Asia & Latin America., 1*(1), 75–92.

López, R. (2019). Proyecto de Lomas de San Miguel es para todos: lmuvi. *PeriodicoCorreo.* https://periodicocorreo.com.mx/proyecto-de-lomas-de-san-miguel-es-para-todos-imuvi/. Accessed 26 July 2019.

Municipal de Cuenca. (n.d.). Registro de la Propiedad: Sistema de Consultas en Linea. *Municpal de Cuenca.* https://www.cuenca.gob.ec/consultas/registro/consultadigital.php. Accessed 30 Nov 2018.

Navarrete, D. (2018). Turismo y gentrificación en ciudades patrimoniales Mexicanas: Exclusiones sociales a través de las transformaciones urbanas y arquitecturales en sitios patrimonio de la humanidad. *Anais Brasileiros de Estudos Turísticos – ABET, 8*(3), 32–46.

News San Miguel. (n.d.). Lomas de San Miguel, la pequeña ciudad donde vivirán 5 mil familias sanmiguelenses. *News San Miguel.* http://newssanmiguel.com.mx/local/lomas-san-miguel-la-pequena-ciudad-donde-viviran-5-mil-familias-sanmiguelenses/. Accessed 26 July 2019.

Pacheco, M. I. F., & Vallejo, M. P. G. (2016). Entre lo local y lo foráneo: Gentrificacion y discriminación en San Miguel de Allende, Guanajuato. *Revista Legislativa de Estudios Sociales y de Opinion Publica, 9*(18), 183–206.

Pallares, A., & Rollins-Castillo, L. J. (2019). Lifestyle migration and the marketization of countries in Latin America. In A. Feldmann, X. Bada, & S. Schütze (Eds.), *New migration patterns in the Americas* (pp. 171–199). Cham: Palgrave Macmillan.

Rainer, G. (2019). Amenity/lifestyle migration to the global south: Driving forces and socio-spatial implications in Latin America. *Third World Quarterly, 40*(7), 1359–1377.

Registros. (n.d.). *Registro de la Propiedad de Cuenca. Registros: Informacion Registral.* https://www.e-registros.es/registro-de-la-propiedad-de-cuenca/. Accessed 30 Nov 2018.

Schafran, A., & Monkkonen, P. (2011). Beyond Chapala and Cancún: Grappling with the impact of American migration to Mexico. *Migraciones Internacionales., 6*(2), 223–258.

Secretaria de Economica. (2016). Regimen de la propiedad inmobiliaria. *Subsecretaria de Competitividad y Normatividad, Direccion General de Inversion Extranjera, Direccion de Asuntos Internacionales y Politicas Publicas.* https://www.gob.mx/cms/uploads/attachment/file/202335/R_gimen_de_la_propiedad_inmobiliaria..pdf. Accessed 29 Nov 2018.

Smith, D. A., Herlihy, P. H., Kelly, J. H., & Viera, A. R. (2009). The certification and privatization of indigenous lands in Mexico. *Journal of Latin American Geography, 8*(2), 175–207.

Steel, G., Klaufus, C. (2010). *Displacement by/for development in two Andean cities.* Paper presented at the 2010 Congress of Latin American Studies Association, Toronto, Canada. https://www.researchgate.net/publication/315542207_Displacement_byfor_development_in_two_Andean_cities. Accessed 26 July 2019.

Terzian, P. (2018, July 10). The world's top 15 cities. *Travel and Leisure.* https://www.travelandleisure.com/worlds-best/cities. Accessed 15 April 2019.

Toone, J. (2018, May 25). The prime of valle de maiz. *San Miguel Times.* http://sanmigueltimes.com/2018/05/the-prime-of-valle-de-maiz/. Accessed 29 July 2019.

United Mexican States Constitution, Article 27, Section 1. https://www.oas.org/juridico/mla/en/mex/en_mex-int-text-const.pdf. Accessed 26 July 2019.

United Nations Educational, Scientific, and Cultural Organization (UNESCO). (2017). *Operational guidelines for the implementation of the world heritage convention.* Paris: UNESCO. https://whc.unesco.org/en/guidelines/. Accessed 11 Nov 2018.

United Nations Educational, Scientific, and Cultural Organization (UNESCO). (n.d.). *World Heritage List.* https://whc.unesco.org/en/list/. Accessed 14 June 2019.

van Noorloos, F. K., & Steel, G. (2016). Lifestyle migration and socio-spatial segregation in the urban(izing) landscapes of Cuenca (Ecuador) and Guanacaste (Costa Rica). *Habitat International, 54*, 50–57.

Williams, A. M., King, R., & Warnes, T. (1997). A place in the sun: International retirement migration from northern to southern Europe. *European Urban and Regional Studies, 4*(2), 115–134.

# Chapter 4
# Social and Cultural Impact of Immigrant Retirees in Cuenca, Ecuador and San Miguel de Allende, Mexico

Lea Efird, Philip D. Sloane, Johanna Silbersack, and Sheryl Zimmerman

Retiree migration from northern, wealthier countries to more southerly destinations with lower living costs and a more pleasant climate has been mushrooming over the past five decades. In the latter decades of the twentieth century, the largest flow of international retirees was from northern Europe to warmer, less costly, countries such as Spain, Italy, Greece, and Malta (Williams et al. 1997). More recently, aging of the baby boomers in the United States and Canada – a cohort that in earlier years tended to push social boundaries – has led to rapid growth of international retirement migration in the Western hemisphere, particularly to destinations in Latin America (Dixon et al. 2006). This generational characteristic of lifestyle innovation, combined with income limitations and financial insecurity, has resulted in increasing interest in "amenity retirement" abroad (Hayes 2015).

Factors contributing to retirement migration include advances in communication (especially internet media such as Facebook, WhatsApp, and Skype), improved international transportation (providing greater familiarity with other countries through tourism), economic globalization (providing more access to familiar products in other countries), and increasing numbers of retirees with incomes that, while substantial in global terms, would allow only a modest lifestyle in their home country (Bernier 2003; Wilson 2013). Migrant retirees are distinct from other migrants and international tourists in a number of ways. Retirees tend to have above-average education and, although their income levels vary, they tend to have consistent income streams from pensions or other sources (Casado-Diaz et al. 2004; Gustafson and Laksfoss Cardozo 2017; Rojas et al. 2014). In contrast, other international migrants tend to be younger and to either seek employment or to escape from a hostile setting (Hayes 2015). Migrant retirees also differ from tourists in that they

L. Efird · P. D. Sloane (✉) · J. Silbersack · S. Zimmerman
University of North Carolina at Chapel Hill, Chapel Hill, NC, USA
e-mail: philip_sloane@med.unc.edu

© Springer Nature Switzerland AG 2020
P. D. Sloane et al. (eds.), *Retirement Migration from the U.S. to Latin American Colonial Cities*, International Perspectives on Aging 27,
https://doi.org/10.1007/978-3-030-33543-4_4

become far more integrated economically and socially into the host country (Bernier 2003).

Migration from one country to another virtually always is associated with challenges accommodating to the new environment. At the same time, if the number of migrants is sizable, the host community has to make its own adjustments. In this chapter we explore how retirees from the United States, Canada, and Western Europe who moved to two colonial cities in Latin America are viewed by natives of host countries, with a focus on the interaction and communication patterns between the two groups. To understand the related issues, we conducted research in two cities – Cuenca, Ecuador and San Miguel de Allende, Mexico. Both are historic colonial cities that are UNESCO World Heritage Sites, with historic city centers that each community has made a commitment to preserve (United Nations Educational, Scientific, and Cultural Organization (UNESCO) 2019). We used a multi-method research approach, gathering data through interviews, internet-based questionnaires, social networking, and observation.

## 4.1   Research Methods

Our primary method of study was to interview local residents, as our main interest was the point of view of the local population in each host city. In both cities, our bilingual research team was tasked with completing 36 interviews, including six interviews with local residents from each of six categories: government officials, realtors, health care providers, non-governmental services providers, convenience store ("tienda") operators in neighborhoods with high concentrations of immigrant retirees, and convenience store operators in neighborhoods with few or no immigrant retirees. We identified the individuals to interview using a combination of networking by local coordinators, internet searching, word of mouth, and (in the case of the tiendas) walking in communities known to have high or low retiree concentrations. All 72 desired interviews were completed, plus a few extras when one or more individuals who had a point of view that the team felt was important to include were identified late in the data collection process; in total, 79 interviews were conducted in Spanish.

The interview questionnaire and additional details regarding the methodology are included in the Appendix. Sections relevant to communication and interaction patterns – the topic of this chapter – included open-ended questions regarding the impact of the retirees on the quality of life of local residents, relationships between local citizens and immigrant retirees, conflicts between the two communities, the perceived integration of retirees into the local community, and perceived positive and negative impact of retirees on the community. Interviews were audio recorded and transcribed for analysis. Analyses were conducted in Spanish and relevant quotes were translated into English with the aid of native speakers. The interview also included some close-ended questions, using a series of Likert-style items.

Analyses used NVivo, a software package for conducting qualitative analyses (https://www.qsrinternational.com/nvivo/home). The analytical coding system is described in the Appendix. Codes applied to the interviews and analyzed in this chapter include (a) all codes related to social dynamics (relationships, harmony, conflict, segregation, integration, paid relationships, power/privilege, philanthropy); (b) communication and communication barriers; (c) cultural similarities and differences; (d) positive and negative characteristics of retirees; (e) statements related to prejudice; and (f) codes related to migration (general attitudes, references to internal migrants, migrants returning to their home country, and international migration from other Latin American countries).

In addition, through observation, field notes, and social interaction with both retirees and local interviewees, the research team viewed and recorded first-hand experiences of the social dynamics between local citizens and immigrant retirees. Also, perspectives from immigrant retirees were obtained by way of an online survey distributed in each city to provide their perspective on these issues; relevant questions included who a respondent's closest friend was, to whom they would turn in times of crisis, who their neighbors are, and whether they are interested in developing friendships with local community members. Details of survey methods and results are provided in the Appendix.

Before presenting the findings, it is important to restate two key points that were noted in the first chapter of this book.

- The data, results, and interpretations reflect the perspectives shared by the people who participated in the study. Because the key aim of the project was to learn the impact of retirees on the local community, the qualitative interviews were with local residents whose relationship with retirees was expected to be in the context of service – government officials, realtors, health care providers, non-governmental services providers, and convenience store operators. Had we instead or additionally conducted interviews with other local residents (e.g., those sitting in a park or doing laundry), the stories we heard may have been different.
- The information obtained is restricted by the research methods. When interpreting qualitative open-ended statements, it is important to be mindful that just because something is not stated does not mean it does not exist. For example, some interviewees may have had a specific opinion about retirees, but purposefully chose to not share it or perhaps did not think to share it. There is also a caveat for close-ended questions that provided Likert-scale responses: it is not known how individuals interpreted the questions, and whether one person's use of the response "strongly agree" was actually markedly different than another person's report of "agree."

Caveats aside, the study's field work provided rich data about local residents' impressions about retirees, communication patterns, transactional relationships, socio-cultural differences and conflict, and social accommodations.

## 4.2 Local Residents' Impressions About Retirees

Social harmony between local residents and retirees in both Cuenca and San Miguel de Allende is a common thread running through the data. There was virtually no open conflict mentioned between the groups, and in analysis "social harmony" was coded far more than "social conflict". Common descriptions of immigrant retirees include such words as "nice," "polite," "respectful," "cultured," and "friendly," with locals usually complimenting retirees' perceived attributes and some of their behaviors, as well as stating that they enrich the local culture. Similar positive but vague attitudes have been noted by Hayes, who, in a study of retirees in Cuenca, posited that a generally receptive attitude on the part of the local population was a combination of respect for the retiree's economic status yet a persistent racism that favored people with light skin (Hayes 2018). This issue is driven home by a 2015 article in The Guardian, entitled "Why are white people expats when the rest of us are immigrants?" It points out that western white people, who are considered "expats" – as opposed to Africans, Arabs, or Asians, who are "immigrants" – are afforded privilege by a racist system (Koutonin 2015).

To further understand attitudes toward immigrant retirees, we asked all interviewees whether they agreed or disagreed with the following statement: "The immigrant retirees are friendly and open in relation to natives of our city" ("*Los inmigrantes jubilados son amistosos y abiertos en relacion on los nativos de nuestra ciudad"*). Across both cities, 81% of respondents agreed or completely agreed with this statement, 15% neither agreed nor disagreed, and only 4% disagreed. Results were nearly identical for the two cities (two interviewees of 38 disagreed with the statement in Cuenca, as did one of 41 in San Miguel de Allende). These results are displayed in Fig. 4.1.

Overall, the majority of interviewees described retirees as educated, genteel, and cultured, and as being respectful of the local culture. One interviewee put it this way: *"They are very respectful of our customs. They are even more enthusiastic about our customs than we are."*

One positive influence of retirees noted by several respondents was their commitment to environmentalism. Some even said that native born residents should adapt some practices from retirees, ranging from picking up their dog's feces to walking more to get from place to place: *"The culture of recycling and care of the environment comes from the foreigner. What I have been able to witness is that the foreigner is careful to not throw papers on the floor. If they go out with their pets, they bring hygiene supplies so the pets don't dirty our gardens and parks."*

Another common theme was the retirees' appreciation for fine art and music and their support of social causes through philanthropy and volunteerism. All of these factors contribute to the general belief that the retirees are a positive addition to the cities, making them more multicultural, contributing to the community's reputation on the world stage as a good place to live, and promoting harmony through giving back to the local communities.

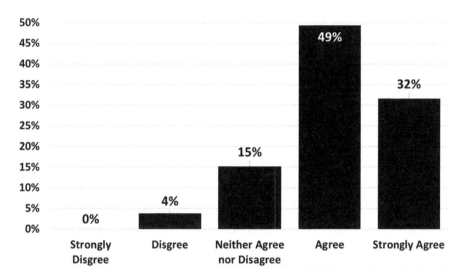

**Fig. 4.1** Responses of a sample of 79 persons from the local populations of Cuenca, Ecuador and San Miguel de Allende to the statement "The immigrant retirees are friendly and open in relation to natives of our city." Responses from the two cities differed little and therefore were combined in this graph. The percent shown is the percent of individuals who chose each of the five response options

Despite these positive sentiments, a lack of integration between the immigrant retirees and the local population was frequently noted. Interviewees pointed to retiree activities and social circles that tended to be separate from those of locals. In addition, retirees were seen to shop differently because of their food habits and preferences, generally favoring supermarkets and large open-air markets over small tiendas.

In both cities, but especially in San Miguel de Allende, the retirees are often physically and/or socially separated from local residents. The creation of American-owned businesses to serve retirees, including restaurants, home care agencies, real estate companies, hairdressers, and laundromats, to name a few, contributes to the creation of a cultural enclave. The interviewees' impression of retirees was that they eat at restaurants much more than local residents, often using these venues for social outings with other retiree friends, to watch American sports, or to celebrate American holidays. These cultural enclaves built around retirees' interests and communities have been found in other retirement destinations as well – for example, a Swedish network of housing in coastal Spain, or English newspapers and football bar nights geared towards Britons' into Didim, Turkey (Gustafson 2016; Nudrali and O'Reilly 2016). Interviewees' also pointed to retirees' attendance at many cultural events, particularly Western music and fine art events, that may not be well-attended by locals. This valuation and importation of Western fine art culture is also one of the social values that locals in both cities identified as different from their cultures.

These points are not meant to convey that retirees did not patronize local establishments or know Spanish; in fact, many interviewees reported that in both cities a significant portion of retirees at least attempt to speak Spanish or attend events with locals. Nonetheless, the tendency of retirees to move within their own enclaves if they so choose speaks to a separateness arising from cultural, economic, and social differences that have allowed them to establish what is, in effect, an American colony within the city. As one interviewee said: *"They stick together. Almost all of them know each other. When we see a new American, we ask them who they know on social media. They have Facebook groups; they have Gringo Post; and they have Cuenca High Life. They learn about everything and find answers to their concerns. They help each other a lot. In that respect the Americans are well-organized."*

One Cuencan resident spoke frankly about concerns that the retirees prefer being separate from intensive involvement with local residents: *"I think that (the retired Americans) aren't really interested in acculturating, in immersing themselves in the culture, or in being part of society. Very few do it, really. Most hang out with other retired Americans, living together, transforming a neighborhood or building into a little American state."* Another put it this way: *"As long as they refuse to integrate and still call Kansas or Arkansas home, they will remain only tourists."* Such responses clearly reflect a negative attitude toward social segregation of the retirees; however not all interviewees felt the same. Some did not see a problem in retirees mostly keeping to themselves, considering it natural because of language and cultural barriers. Thus, there are varying points of view about retirees' physical and social separation, although the majority of local interviewees thought that more integration-focused activities and social events would improve relationships between retirees and local residents.

Other interviewee comments focused on differing cultural norms between the retirees and local population, including punctuality, the speed of governmental systems, and professional formality, which Americans emphasize much more than Latin Americans, as well as the differences in how people use physical contact in greeting each other. Interviewees often reported that retirees expected Latin American culture to change to accommodate them, rather than they themselves adjusting to the local cultural norms. However, one interviewee did say that some retirees integrate: *"Many already are quite integrated. But some have a really hard time, and it's like a culture shock. I would say that 70% integrate well, but the rest have trouble, especially because of the way things work in this country – much slower and in a different, more personal, manner."*

Interviewees consistently mentioned philanthropy and volunteerism as a contributor to their positive opinion about retirees' presence in their cities, and the fact that these activities enrich and support the local community in important ways. "Giving back" appears to be a common element of retirement migration worldwide, providing multiple benefits not just to the communities but to the retirees, including a sense of purpose and a productive social outlet (Haas 2013). Indeed, prior research has found that philanthropic efforts abroad, within a number of international retirement destinations, are a major source of relationship and social capital building within retiree circles (Casado-Diaz 2016; Croucher 2011; Foulds 2014).

The extent of charitable giving and volunteerism seen in migrant retirees was noteworthy because it reflected some cultural differences. In individualist societies such as the United States, robust nongovernmental charities and volunteerism have a key role in enhancing the safety net for poor, underprivileged, old, and disabled persons and in providing disaster relief. Compared to the United States, nongovernmental organizations (NGOs) are a relatively recent phenomenon in Latin America. Modern NGOs in Latin America began in earnest in the second half of the twentieth century, primarily driven by the Catholic Church. Since that time, they have experienced periods of instability (Valdivia 2014), and so their presence is not as established as in the United States. Still, substantial non-profit work is being undertaken by the local communities, which in some cases has been overshadowed by the media's portrayal of foreigners' contributions (Covert 2010).

Our field notes provided the following information about some of the charity-related activities in the two cities:

- In San Miguel de Allende, where retiree-linked charitable organizations are so numerous as to be considered an important part of the social safety net (Covert 2017), we attended a quarterly event entitled "100 Women Who Care". On the night we attended, the room was packed, mostly with retirees. The gathering's purpose was to award a 100,000-peso grant (approximately $5250) to a charitable organization. Twenty-five charities had entered their names in a hat; three names were drawn; and a representative from each had 5 minutes to speak to the group in English about why their organization should win and what they would do with the money. The winning charity was a recently reopened Hospice.
- In Cuenca, an event had occurred the previous month in which the governor of the state of Azuay recognized nine charitable organizations headed by retirees for having contributed to "saving lives and relieving suffering" in the community. Among the issues addressed by the charities recognized in that event were care of older persons, music concerts for children, a home for persons with AIDS, a soup kitchen, and a charity providing aid to abandoned animals. The winning charity was a refuge for women who experienced domestic violence, which had been founded 5 years earlier by a retired American (de Azuay 2018).

Not all interviewees found this philanthropy positive. One commented that the retirees tend to pick causes in which they are interested, and not necessarily what is most needed by the community – a paternalistic pattern of giving seen in other retirement destinations within Central and Latin America (Croucher 2011; Foulds 2014; Covert 2010; Benson 2013). Another noted that charitable organizations can cause the government to pull back on services, noting that, because a retiree-supported NGO in San Miguel de Allende was building wells in outlying villages, the state government-initiated cuts in its own local infrastructure funding.

A greater individualism among retirees, contrasted with a more collective focus in Latin America, was also perceived and was felt to extend to perceptions of family. Retirees are felt to focus more on the individual and the nuclear family, whereas local residents tend to consider family to be a multigenerational extended network. In other research, some retirees in San Miguel de Allende and Ajijic, Mexico went so far as to attribute this differing familial value as a cause for the segregation between the groups. The research found that a third of those interviewed felt that "socializing in Mexico takes place primarily in the family and that unless you marry into the Mexican family you are not likely to be truly embraced within the inner circle." (Croucher 2011) From the perspective of the local population, even the idea of moving far away from family, particularly children, in order to retire (as opposed to a "necessary" move to secure employment or escape difficult circumstances) is counterintuitive to their view of family (Nudrali and O'Reilly 2016). Such an individual focus also may differentiate migrant retirees from many of their peers and their own cultural roots (Huber and O'Reilly 2004).

## 4.3 Communication Patterns Between Residents and Retirees

Although cultural differences were evident within the interviews, language was by far the most commonly identified barrier to integration noted by local residents. This is not a new finding – segregation due to language is a theme that has been emphasized by previous research (Gustafson 2016; Casado-Diaz 2016; Croucher 2011; Hayes and Carlson 2018). Within our research, the topic was especially common in the interviews conducted with Cuenca residents, one of whom stated: *"We have American friends because we speak English, but 90% of the (native) people don't. It's a barrier to relationships. When we go to gatherings of Americans, there are hardly any Ecuadorians."* This quote highlights the crucial nature of language in creating more integrated social circles, something that was reported as lacking in both cities. As one interviewee stated: *"Many come with the intention of getting involved with the local community; however, they [the retirees] don't succeed, because of the barriers of language and cultural differences."*

At the heart of this communication barrier is perceived to be the relatively poor Spanish language skills of the immigrant retirees. As one interviewee said: *"Their language is elementary. They speak a little with the cook, the gardener, the taxi driver, and the storekeeper. They don't need to speak Spanish."* Although some retirees attempt to learn Spanish and there are Spanish classes available in both cities, local residents postulate that few retirees become proficient because of a combination of their advanced age, the availability of local residents who are bilingual, a tendency of the retirees to socialize together, and a lack of motivation on behalf of the retirees to speak Spanish due to the prevalence of basic English in both cities.

## 4.4 Transactional Relationships Between Immigrant Retirees and Local Residents

As reported above, local residents we interviewed tended to speak in generalities about their interactions with or impressions of immigrant retirees. Interviewees tended to speak about them as "nice," "good people," or "polite" – general terms that were often accompanied by statements indicating limited personal contact with individual retirees. What interactions were reported tended to be brief and casual, as with a neighbor or customer, perhaps complicated by the significant language and cultural barriers previously discussed. In fact, only eight of the 79 interviewees (10%) mentioned a sincere, non-paid friendship with at least one retiree friend or neighbor. One interviewee phrased the reality as: *"For the majority of people, it's a null relationship. A relationship doesn't exist, except the eventual commercial relationship. I've worked with many Americans, yet I have very few American friends."* In the words of another interviewee: *"[The relationship between the communities] is more like a business plan. We have to be nice [to them]. We are people with big hearts and want them to feel comfortable being here in the city. The relationship is more obliging ["servicial"] than friendly."*

Part of the lack of strong friendships may be attributed to the power dynamics that occur when the majority of relationships are built around monetary transactions. Research has found that the purchasing power differential due to vastly different socioeconomic statuses has made some local residents in Didim, Turkey feel second-class compared to the incoming British migrants (Nudrali and O'Reilly 2016). Although this attitude was not held by the majority, a quarter of the interviewees in San Miguel de Allende agreed with the statement, "the presence of retired immigrants makes me feel like a second-class citizen in my own city". Conversely, only 8% of the Cuencanos interviewed endorsed this statement – not a high percent, but still present.

Lack of close relationships is likely representative in general of the local residents of San Miguel de Allende and Cuenca; however, we stress that our interview strategy excluded local residents who were employed part or full-time by retirees. Between a third and half of interviewees brought up the subject of retirees hiring local workers as domestic help, and these paid relationships clearly also shape the public consciousness about retiree-local relationships.

Being employed by a retired American brings financial benefit, as the Americans have a reputation for paying better than the norm in both cities. As one local resident stated: *"Many foreigners hire service personnel such as gardeners, housekeepers, nannies, et cetera. And they pay much better than a regular job, and they pay by the hour. Yes, they have a positive impact on many local families because they get a better income, have better benefits, and have a better work situation."* Additionally, these transactional relationships can be a way of bringing neighbors together, promoting integration between retirees and the local population. One interviewee explained it this way: *"There are neighborhoods where they integrate (with local residents). For example, their neighbor next door or in front of their house may be the person who helps out with cleaning."*

The benefits of working for a foreign retiree are not, however, perceived to have a ripple effect on the broader community, except possibly to increase prices. This point was expressed by one interviewee as follows: *"The people from San Miguel who work with foreigners are better paid, and this improves their quality of life. Only those who work for the foreigners; the rest see no benefit at all."*

These transactional relationships, while structurally not between friends or social equals, can lead to strong bonds. Research has found that these relationships can go above and beyond a typical professional dynamic, and result in local families "adopting" the migrant retirees they work for (Hayes 2018). It is not unusual, for example, for retired Americans to report getting to know a housekeeper's entire family and being invited to family events. It also is not unusual to hear stories of retirees helping employee's families, for example by paying school fees or buying school uniforms, or even helping look after the children of a single mother who works in their home.

The following description from our field notes illustrates the closeness that can develop in such relationships.

*"[The Housekeeper] doesn't speak much English but is still able to communicate with [the retired American] and her friends. [The housekeeper] clearly views [the retiree] as part of her family and feels very attached to her. When asked about her relationship with [the retiree], she started to tear up, expressing 'La quiero mucho mucho mucho. Es como mi abuelita, mi mama [translation: I love her very much. It's like my granny, my mom]'. To which [the retiree] responded 'I wish she wouldn't say that, I wish she would say I was more like her sister', which made both of them laugh. They seemed really comfortable with each other despite their power dynamic."*

This interchange illustrates how close these transactional relationships can become, and yet that they also reflect the general social distance between the two groups, which is perpetuated despite their frequent transactional social contacts.

Another source of transactional relationships is native residents who have learned English, often by spending time in the United States, and who connect to retirees through businesses such as taxi driving. Indeed, both cities had a few individuals who, by virtue of being bilingual and educated, offered their services as "facilitators" – serving as mediators between non-Spanish speaking retirees and local agencies or businesses. Additionally, because these returning migrants know English well and share many aspects of North American culture with the retirees, these relationships can more easily become peer-like.

Facilitators differ from translators in that they do more for the retirees than simply interpret. Instead, they act as cultural liaisons, helping retirees do everything from negotiate in markets or with taxi drivers to rent a home or arrange a visa. Their work is invaluable to newly-arrived retirees in particular, for whom facilitators can provide guidance on the city's neighborhoods, culture and resources. Facilitators are one element of what is being dubbed as the 'migration industry', a segment of the market specifically geared towards providing services, comforts, and experiences for expats (David et al. 2015). The following advertisement from a Cuenca expat website illustrates how facilitators present themselves and what they do:

> *"My name is XXXX. I am an Ecuadorian, but I lived in Canada for many years and speak English. I have been a Facilitator for 5 years. I have assisted dozens of expats in obtaining their visas, cedulas [translation: government documents], health insurance, housing and much more. I also assisted expats wishing to become Ecuadorian citizens. These services are available as one fee package or individually. I own a car, so transportation is always included at no extra charge. I have always gotten my clients their visas, even under the most difficult of circumstances, such as unreadable fingerprints for FBI reports. I offer free consultations, either by telephone, email or in person. I am at the immigration office on a very regular basis, so I am knowledgeable about the new immigration law and am happy to share my knowledge with you. I will be happy to provide you with references. Please feel free to contact me for a free consultation."*

## 4.5  Socio-cultural Differences and Conflicts

Interviewees rarely reported open conflict between immigrant retirees and the local population. However, many cultural differences were identified by locals that can affect relationships between the two groups. These differences can be broadly themed into norms about *interpersonal behavior*, norms about *family*, and norms involving *social and cultural events*.

Interpersonal behavior in casual social settings emerged from our interviews as an important norm for locals that retirees do not often follow. For example, locals mentioned that immigrant retirees did not generally use the same type of physical contact in greeting people that locals use; they also reported that retirees often get frustrated with cultural differences around punctuality, in that Latin American culture sees lateness as a norm but retirees may see it as rude. Similarly, in Latin American culture, all relationships, professional or not, are imbued with relaxed, friendly conversation; retirees were said to be more formal and timelier and to often expect the local population to comply with their cultural norms.

Another interpersonal cultural difference is that retirees do not engage in the small talk that is perceived as polite in San Miguel and Cuenca, including with their neighbors and within transactional relationships. One local interviewee said: *"I used to live in an apartment, and for four or five years my next-door neighbors were an American couple. We talked once or twice but never socialized."* This comment reflects a social convention that one should know and be friendly with one's neighbors, which is not necessarily true in retirees' cultures of origin. Another issue is a perception that many retired Americans do not understand the Latin American custom of valuing small talk and relationship-building before beginning any transaction, including such business interchanges as making a purchase in a store or having a workman do repairs in an apartment. This point was illustrated by a discussion one of the research team had with staff in a store in Cuenca: *"The women were all in agreement that immigrant retirees do not spend their money in this store ... and talked about how many of them walk in and right back out, without any greeting or acknowledgement about the people running the stores. The storeowner emphasized that it was the duty of the person entering a place of business to greet people in the store. Another woman chimed in that even though she doesn't know much English, she knows how to say hello."*

Another theme raised in multiple interviews is the strength of family ties. The concept of leaving one's family to retire far away seemed difficult to understand, especially because it separated aging parents from their children. As one interviewee stated: *"It is my understanding, more or less, that over there a young person turns 18 and leaves the house. They don't return to see their parents for a while – a year, two, three, five years, ten years. The parents only receive calls on Mother's Day or Christmas. We don't do that here....We have different customs. The concept of a nuclear family is stronger here, I would say."* In contrast to this perceived characteristic of retirees, local interviewees in Cuenca and San Miguel de Allende expect their children to leave home later, and multigenerational households are often the norm, especially when aging parents begin to need assistance. Family time is also an important concept for local residents, in that non-work hours and weekends are strictly reserved for personal and family time. This cultural difference is also echoed in local interviewee comments on migrants returning to their home countries, in that one of the negative consequences of economic migration to richer countries was the family separation that resulted, and a positive result of return migration is the ability for families to reunite.

Choice of social activities was another social difference identified by interviewees. Local residents noted that retirees congregated in cafes and restaurants and often visited museums, concerts, and festivals. These events were not the same as those that local residents attended or valued, of which family time and city-wide religious and cultural holidays were especially important. In both cities, local interviewees spoke about the importance of these holiday events, including parades and religious activities, and suggested that retirees not just take photographs but join as participants. In addition, a few interviewees mentioned that retirees use substances at higher rates than locals, for example, drinking large quantities of alcohol at once, and one interviewee expressed great concern about them using marijuana.

The few areas of open social conflict reported by interviewees centered around noise complaints, language barriers, and treatment of service sector workers. Many local interviewees stated that retirees would have disagreements with their neighbors over the playing of loud music at local fiestas or noise generated from cultural events in the city center, such as the frequent fireworks in San Miguel de Allende. This seems to stem from retirees not communicating well with their neighbors or not understanding the number and importance of religious, political, and social holidays that are celebrated in both cities. Language, not surprisingly, was mentioned frequently, as it can cause misunderstandings or make communication generally difficult.

The strongest reports of conflict reported by interviewees involved stories of individual retirees yelling at or being rude to service staff. Such incidents were reported in both study cities and ranged from retirees yelling at waiters and cashiers for not speaking English, to refusing to put out a cigarette in a non-smoking area, to not understanding the Ecuadorian concept of "la yapa" (an optional, extra gift that a store owner can choose to give customers, such as the inclusion of an extra baked good to make a baker's dozen) to trying to haggle over fixed-price items. Twelve of our interviewees across the two cities reported witnessing these types of interactions.

## 4.6    Social Accommodations for Retirees

As the numbers of retirees have grown in Cuenca and San Miguel de Allende, the government and business communities in both cities have made adaptations to accommodate retiree wants and needs. In both cases, immigrant retirees were seen as good for the city. *"I think that we have opened our doors to them,"* one respondent said. *"We have given them space, including the authorities, who opened up spaces for them to work, to integrate, and to have a mutual dialogue."* This willingness to include retirees in the public life of the cities was echoed by multiple interviewees, who highlighted social and governmental actions that have been aimed at including the retirees in their society and culture.

A key accommodation has been an emerging bilingualism in the service and government sectors, as local residents working in these fields have an economic incentive to learn English. Thus, as noted in Fig. 4.2, local residents we interviewed strongly endorsed the idea that learning English was essential to be successful in

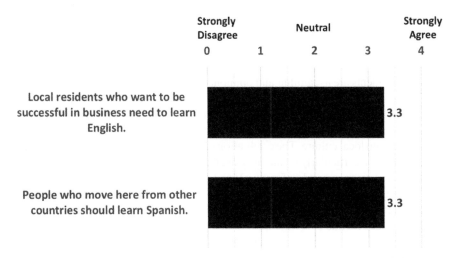

**Fig. 4.2** Responses of a sample of 79 persons from the local populations of Cuenca, Ecuador and San Miguel de Allende to statements regarding local residents learning English and immigrant retirees learning Spanish, using a scale ranging from 0 = strongly disagree to 4 = strongly agree. Responses from the two cities differed minimally and therefore were combined in this graph. The number shown is the mean score for each question

business (scoring it 3.3 on a scale of 0–4). Additionally, interviewees endorsed with similar enthusiasm (score 3.3) the idea that immigrants from other countries should learn Spanish, as is also noted in Fig. 4.2.

In both cities, the governmental sector made additional accommodations for immigrant retirees. Both cities have a US consulate, both created government outreach programs for the immigrant retirees, and both sponsor public events for the retirees involved in city life. One interviewee reported that the government in San Miguel de Allende goes even farther, giving retirees serious consideration when discussing potential policies around themes like security and public services, and encouraging their involvement and/or input in government processes. San Miguel de Allende also has a bilingual Public Ministry, specifically so that retirees can receive assistance – including with reporting crimes – in their native language.

## 4.7 Key Differences Between Cuenca and San Miguel de Allende

Although many of the same social dynamics were reported in the data from both Cuenca and San Miguel de Allende, there are some important differences in the cities' adjustment to the immigration of retirees. Overall, there is a theme of the cities being in differing stages of immigration. Specifically, Cuenca, which has a shorter history of retirees living there and smaller numbers of retirees, is in an earlier stage of adjustment. San Miguel de Allende began seeing Americans in significant numbers as early as the late 1940's and has been a retiree destination for several decades; not surprisingly, its population at 10% "gringo" is proportionally much greater than that of Cuenca, where retirees constitute only around 2% of the population. Additionally, Cuenca is a commercial center of the southern Andean region of Ecuador and is nearly five times as large as San Miguel; so, it is perhaps surprising that local residents are as familiar and opinionated regarding the retiree migration as they were.

In San Miguel de Allende many local residents we interviewed talked about cultural fusion of holidays when asked about how retirees were integrating, or not integrating, into local culture. The celebration of the Fourth of July and Thanksgiving in retiree spaces, particularly restaurants and bars popular with or owned by retirees, was identified as an example of cultural integration into the city. The most salient example was the fusion of Halloween and the Day of the Dead, two similarly timed holidays, although they have different cultural meanings. Many local residents in San Miguel de Allende celebrate both holidays simultaneously, having picked up the decorations and trick-or treat traditions from retirees and other immigrants over the past decades. This cultural fusion was not reported in Cuenca, most likely due to the fact that the retirees' presence is not yet ubiquitous enough to warrant the merging of cultural traditions.

Another key difference between the two cities, which tangentially affects opinions about retirees, is the fact that the retirees are only part of the transcultural mix in each city, and the other components of that mix differ and impact each city in unique ways.

In Cuenca, two other types of immigrants are particularly important:

- Latin American immigrants, who are largely refugees from violent and/or economically depressed settings. In previous years they came largely from Columbia, but lately they have been overshadowed by an influx fleeing the chaos in Venezuela. These immigrants were regarded with suspicion and xenophobic sentiment by some interviewees, with one respondent saying, *"We are in the middle of a Venezuela problem. The Venezuelans have invaded us these past few years."* Interviewees attributed organized crime and drugs, loss of jobs, and an overall negative economic impact to the presence of these immigrants. This perspective contrasts with the prevailing opinion that retirees from North America and Europe provide economic benefit and reflect valued cultural attitudes and behaviors.

- Return migrants, who left Ecuador during bad economic times and are now both returning in large numbers and contributing economically by sending remittances to families and investing in property (Herrera and Martinez 2015). Emigration was particularly intense during the crisis in the late 1990s, when nearly a million Ecuadorians, many from Azuay province (where Cuenca is the capital) emigrated to other countries. The main destinations of this immigration were Spain, where they tended to immigrate legally, and the United States, where they tended to not obtain legal residency. As of 2013 an estimated 428,500 Ecuadorians lived in the United States and 456,233 lived in Spain (Jokisch 2014). This immigration flow has been reversed in recent years, due to improvement of the Ecuadorian economy, hard economic times in Spain, and increasing pressure on undocumented workers in the United States. These returning migrants vastly outnumber the American retirees, and many have returned with social capital (language skills, business skills) and financial capital (in one study, returning migrants brought with them an average savings of $40,000) (Grunenfelder-Elliker 2011). Indeed, many of the daily services used by retirees from the United States are provided by returning Ecuadorians who speak English and understand the preferences of Americans (Herrera and Martinez 2015). These services include domestic help, translators, facilitators, and drivers. Many businesses frequented by "Gringos," such as real estate offices, cafes, and restaurants, are owned and/or operated by returning migrants.

In San Miguel de Allende, the "outsiders" about whom we heard repeatedly were different. There, too, we identified two predominant groups:

- International and Mexican tourists. In recent years San Miguel de Allende has become known as a highly desirable tourist spot. It was selected in both 2017 and 2018 by readers of Travel & Leisure magazine as "the world's best tourist city." (Terzian 2018) It also has become quite popular as a destination wedding site, having hosted a reputed 500 destination weddings in 2017. Tourist cars and buses crowd the downtown area, especially on weekends, and the streets are often filled with rowdy tourists late into the night. Furthermore, purchase of homes in the city center for conversion into hotels or short-term rentals through Airbnb or VRBO has been attributed by interviewees with driving up real estate costs more than have the immigrant retirees.

- Mexicans who have purchased second homes or retirement homes in the city. Because of its reputation for beauty, restaurants, shops, and safety, San Miguel de Allende has become quite popular as a second home or retirement destination for wealthy families from Mexico City, Querétaro, and other cities. For the same reasons it has also, according to some of our interviewees, become a "safe" place for gang leaders to purchase homes and locate their own families. As far as we could tell from our interviews, the influence of these part- and full-time migrants from other areas of Mexico has been mainly financial rather than cultural.

## 4.8   Suggestions for Increasing the Harmony Between Retirees and the Local Population

As part of the interview protocol, local residents were asked for recommendations to improve relationships between the two groups, and several common themes were expressed. One theme was a wish for the retirees to express more sensitivity to and adapt better to local cultural norms. As one interviewee stated, *"The retirees need to understand how we do things here. I think it is good that they move here, but they should accept our rules and procedures, and they should treat Ecuadorians as they want to be treated."* In particular, several local residents wanted retirees to accept that they spend time with their families during certain events (around Christmas, for example) or times of the week, that family takes a central role in their lives, and that such patterns were legitimate reasons to not hold a meeting or respond to a service request.

Interviewees also suggested that the retirees learn more about and participate more fully in cultural events, like Corpus Cristi, Semana Santa, Dia de los Muertos, and el Paso del Niño. They also thought that cultural exchange – retirees sharing their customs, as well as locals sharing theirs – would be a good way to bring more unity and understanding of each other. One interviewee suggested that food could function as a tool for this cultural exchange, saying that a fair with foods from different countries of origin, as well as local food, could be an event that aided integration.

Also, as noted previously, one of the most commonly mentioned recommendations for integration was reducing the language barrier by having retirees learn more Spanish and local residents learn English. Interviewees thought that improved language skills would help retirees learn more about local culture and thus improve relationships with local residents. One interviewee even suggested a program to assist with this goal that would involve pairing retirees with schoolchildren to do a language exchange.

## 4.9   Retirees' Impressions of Communication and Social Dynamics

To gauge whether and to what extent the opinions of immigrant retirees themselves about communication and social dynamics are similar to those of the local population, we are able to draw from responses to our online survey, plus comments in our field notes and a few formal interviews with retirees themselves. These data validate the opinions stated by local residents about language skills, in that over 70% of survey respondents in both cities rated their own ability to speak Spanish as at best limited to simple conversation (Fig. 4.3). These results are quite similar to those identified in a survey conducted several years ago in San Miguel de Allende and the Lake Chapala region of Mexico by Rojas et al. (Rojas et al. 2014)

Despite their limited ability to speak the native language, the majority of immigrant retirees who responded to the survey expressed interest in developing friendships with people who are part of the native community, with the majority neither agreeing nor disagreeing with a statement that they were more comfortable spending time with people from their own culture (a mean score of 3.2 on a scale of 0–6, where 3 = neither agree nor disagree; see Fig. 4.4). These results contrast with the opinions expressed by the representatives of the local population we interviewed, who, as noted earlier, tended to see the retirees as socializing with each other and remaining relatively separate from the native community. This general feeling of

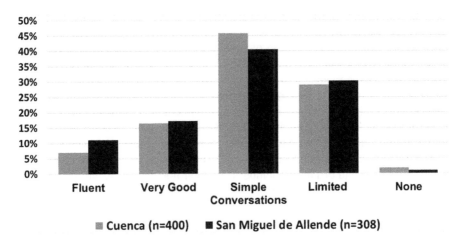

**Fig. 4.3** How immigrant retirees in Cuenca (N = 400) and San Miguel de Allende (N = 308) rated their own ability to speak Spanish. The percent shown is the percent of individuals who chose each of the five response options

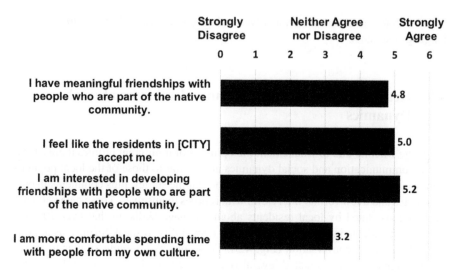

**Fig. 4.4** Average responses of immigrant retirees in Cuenca and San Miguel de Allende to statements about social relationships with the local population, using a scale ranging from 0 = strongly disagree to 6 = strongly agree (N = 659–671 depending on the item). Responses from the two cities differed little and therefore were combined in this graph. The number shown is the mean score for each question

integration is not atypical for migrant retirees. A previous study of retirees in San Miguel de Allende and the Lake Chapala region of Mexico found that 99% of respondents to a survey felt "moderately well-adapted to the Mexican culture." (Rojas et al. 2014) Given the relative lack of language skills and acculturation, this attitude could be considered to reflect an "imaginary" idealization more than a reality (Hayes 2018), perhaps reinforced by the very real feeling that the retirees are welcome and accepted by the host city.

When we examine the response of immigrant retirees to questions about who they live adjacent to, who their closest friend is, and what they would do in an emergency, we see some evidence of integration with the local community. Although closest friends do tend to be largely immigrant retirees like themselves, 29% in Cuenca and 19% in San Miguel de Allende identify their closest friend to be a native of the host city. This integration between the cultures is particularly evident in responses to the question "Who would you turn to if you needed somewhere to stay temporarily because your home had flooded?" Somewhat over half in both countries would turn to a friend or family member from their home country, but most of the remainder would depend on natives of the host community (21–32%). These results are displayed in Table 4.1.

Also noteworthy in the responses of immigrant retirees to our online survey was general agreement to the statement "I have meaningful friendships with people who are part of the native community" (mean score 4.8 on a scale of 6; Fig. 4.4). Based

**Table 4.1** Responses of immigrant retirees in Cuenca, Ecuador and San Miguel de Allende, Mexico, to questions about their nearest neighbors and closest friends

| | Cuenca | San Miguel de Allende | Overall |
|---|---|---|---|
| **My closest friend is a native of:** | | | |
| United States/Canada | 62% | 75% | 68% |
| Study city (San Miguel de Allende or Cuenca) | 29% | 19% | 24% |
| Other | 9% | 7% | 8% |
| **Who would you turn to if you need somewhere to stay temporarily because your home had flooded?** | | | |
| Family member/friend who lives with you | 7% | 7% | 7% |
| Person from home country | 47% | 54% | 51% |
| Person from local community | 32% | 21% | 27% |
| Other | 14% | 18% | 16% |
| **Where is your [first] closest neighbor from (born or spent much of their life)?** | | | |
| Study city (San Miguel de Allende or Cuenca) | 69% | 56% | 63% |
| USA/Canada | 25% | 39% | 31% |
| Other | 4% | 4% | 4% |
| Don't know | 3% | 2% | 2% |
| **Where is your [second] closest neighbor from (born or spent much of their life)?** | | | |
| Study city (San Miguel de Allende or Cuenca) | 69% | 53% | 62% |
| USA/Canada | 23% | 39% | 30% |
| Other | 2% | 5% | 3% |
| Don't know | 6% | 4% | 5% |

Note: Responses varied depending on the question, from 400 to 408 respondents in Cuenca, and from 305 to 314 respondents in San Miguel de Allende

on both our interviews and observations, it appears that most of these friendships are transactional in some way, frequently beginning and often continuing on the basis of a paid relationship – for example, with a housekeeper or favorite taxi driver. Immigrant retirees with better Spanish language skills tend to more commonly report close friendships with local residents, an observation that is modestly supported by a Spearman correlation coefficient of .19 ($p < 0.001$) between self-reported ability to speak Spanish and having "meaningful friendships with people who are part of the native community." Interviews with retired Americans we encountered reinforce these data.

Although many relationships were at the basis transactional, interviewees rarely discussed the monetary aspect of these friendships. Whether this omission reflects an acceptance of the relationship's value because of the feelings engendered despite its transactional nature, or whether awareness of the transactional nature of the friendship has faded with time is something we were unable to explore, but reflects an interesting dynamic of the retiree-local interactions.

## 4.10   Discussion

Our interviews with local residents, combined with survey data from retired immigrants in the two study cities, indicate that retiree immigration to Cuenca and San Miguel de Allende has had noticeable but not overwhelming effects on the cities' culture and social dynamics. Although there is an overarching theme of harmonious coexistence between local residents and immigrant retirees, there are deeper, more complex realities at work – some of which suggest that it may require purposeful effort for the host cities to maintain their identity if retirees become more active but not fully acclimated in the culture.

Local residents almost always spoke in generalities about the retirees, praising their behavior and contribution to social causes, but rarely speaking of deep friendships. These observations indicate a relative separation of the retiree community from local society, with language identified as a major cause of this lack of integration; however, such separation does not threaten the identity of the host community. One wonders, though, about what may happen if the retirees do as many of our local interviewees desire – learn the language and better integrate in the city. Such integration must fully respect the local culture to not change it. These and other themes provide a complex picture of a dynamic process.

Natives of both cities overwhelmingly felt positive about the presence of the immigrant retirees. They reported the retirees as a group to be an economic boost to the community, and as individuals to be friendly, respectful of local customs, and environmentally conscious. The finding that local residents found the economic impact of retiree migration to be positive is consistent with research in other settings (Serow 2003). Still, such a point of view, while accurate in the short run, may downplay long-term negative impacts, such as environmental degradation and disruption of local social structures (Serow 2003; Barrantes-Reynolds 2011). Indeed, a study using similar methods to ours interviewed 45 local residents in three "mature" international retirement destinations in Spain about their perception of the impact of retirement migration. Conclusions included that the migration vastly improved the economy, helping bring about "modernization," but at the expense of the community's cultural identity and degradation of the environment (Huete et al. 2007).

Among the retiree behaviors looked upon with favor by many interviewees was involvement in philanthropic and charitable activities. Our interviews, as well as comments from retirees who responded to our online surveys, indicate that volunteering and support of charitable causes are common and comprise a key element of bridge-building between retirees and the local community. Academics have been more critical of this practice, however, suggesting that philanthropic activities could be considered a form of "neocolonialism." (Benson 2013) Indeed Covert, in her book on San Miguel de Allende, cynically referred to non-profit and volunteer work within the Mexican community as a "hobby," (Covert 2017) and Hayes, in his book on retirees in Cuenca, referred to charitable work as "a contemporary continuation of settler forms of colonialism." (Hayes 2018) Benson, in research on North American retirees to rural Boquete, Panama, concluded that volunteering, charitable work, and philanthropy by the economically better-off retirees did not address the underlying social disparities but rather served as an expression of privilege and power (Benson 2013). These comments are in line with those of a few of our interviewees, who, as earlier noted, raised questions about the value of retiree charitable activities. On the other hand, recognizing the complexity of the issue, it can be argued that these criticisms are not entirely justified. Sensitized to community needs, retirees can be seen as having responded by making well-intentioned contributions to the welfare of their adopted communities – efforts that were largely recognized by the majority of our interviewees.

Despite these and other visible activities, local residents perceived the retirees as largely keeping to themselves. Social separation of retirees from local residents is created and maintained by retirees having separate circles in which they conduct business and have relationships. Eating, socializing, and attending events that are different from the local population, as well as not adjusting well to Latin American cultural norms, creates and maintains social separation and is often perceived by locals as an undesirable aspect of retiree immigration. What is unclear is whether more social integration is a reasonable expectation. Multiple studies in other settings have demonstrated that it is common for international retirees to surround themselves with persons of similar origin and personal characteristics and to prefer locations that facilitate such relationships, such as restaurants (Dixon et al. 2006; Bernier 2003; Gustafson 2016; Nudrali and O'Reilly 2016; Huete et al. 2007). Further, cultural differences such as religion and family closeness have, for example, created divisions between British expatriates and local residents in Didm, Turkey (Nudrali and O'Reilly 2016), and between local residents and Swedish immigrants in Spain (Gustafson 2016).

Language was perceived by our interviewees as a major barrier to increased social integration, and many expressed a desire to see more retirees develop better Spanish language skills. In our retiree survey, the majority admitted to no more than basic Spanish, a finding consistent with prior research (Rojas et al. 2014). Not learning the local language has been similarly observed among long-established retiree communities in Spain, with reasons for not learning Spanish including the difficulty of language acquisition and the tendency of retirees to spend most of their time with people who speak their own language (Gustafson and Laksfoss Cardozo 2017). The lack of language skills drives retirees to preferentially frequent service providers who have learned the retiree's language, thereby encouraging multilingualism among the host country's labor force (Gustafson and Laksfoss Cardozo 2017).

The connection between knowing the local language and developing close social relationships with the native population is both self-evident and supported by research (Huber and O'Reilly 2004). What is unclear is what degree of language acquisition is realistic for immigrant retirees to obtain, as the level of fluency needed to carry on more than transactional conversations requires many thousands of hours of study and immersion, and is particularly difficult for older adults to master (Hartshorne et al. 2018). Because of the English languages' role as the "lingua franca" worldwide (Gustafson and Laksfoss Cardozo 2017), the economic advantages of learning English for local residents is clear within business transactions –

for example, a recent study of taxi-drivers in Cuenca found that they considered proficiency in English to be an advantage in their line of work, and many voiced an interest in lessons (López 2017). For immigrant retirees, on the other hand, there is little economic incentive to learn more than transactional Spanish. Thus, it seems unrealistic for international retirees to be expected as a group to gain fluency in a new language, and that instead both populations should develop some familiarity with the other's language in order to conduct the typical transactions of day-to-day living. However, for true social integration and assimilation with the local population, making a concentrated effort to learn the host countries' language is key (Gustafson and Laksfoss Cardozo 2017).

In spite of language limitations, many immigrant retirees reported strong relationships with local residents, most of which had a transactional origin that progressed to develop warmth and mutual respect. These transactional relationships explain both the generalizations of interview respondents about retirees and the unwillingness of local residents to confront them if there happens to be a conflict. Some non-transactional friendships do, however, develop, particularly between retirees and local residents who both share high-level language skills in either English or Spanish, and with whom the power dynamics associated with economic disparities are absent or attenuated (Hayes and Carlson 2018).

Another key finding from our interviews is that government and business communities in both cities have made adaptations to accommodate retiree wants and needs. A key adaptation was introduced previously – development of English language skills by persons engaged in business. In a similar vein, the local governments of both Cuenca and San Miguel de Allende developed offices and resources with English-speaking personnel, aimed at assisting international retirees and tourists. In Cuenca, the municipal government has worked hard to make public spaces available to retirees and put resources into an international relations department to provide dictionaries and other resources to help with integration (Álvarez et al. 2017). In San Miguel de Allende, promotion of tourism-friendly events, offices, regulations, and resources has been a strategy for decades, and again, here we note that it has been argued that local government prioritized the growth potential of tourism and immigration over preservation of quality of life for the native population (Covert 2017).

Finally, it must be acknowledged that gauging the impact of retirees on social systems (as well as on the economies) of the two cities is challenging because they are but one of many forces operating simultaneously in each country. In San Miguel de Allende, tourism is at least as powerful a force for social change as is the retiree community. In Cuenca, retirees, while quite visible, share immigrant status with returning Ecuadorians who have been out of the country for years, and with political and economic refugees from Venezuela and other nearby countries. In this context, the international retirees are less socially disruptive than some of these other phenomena, which no doubt adds to the generally favorable opinions expressed of them by local residents.

# References

Álvarez, M. G., Guerrero, P. O., & Herrera, L. P. (2017). *Estudio sobre los impactos socio-económicos en Cuenca de la migración residencial de norteamericanos y europeos: Aportes para una convivencia armónica local* (Informe Final). Cuenca: Avance Consultora.

Barrantes-Reynolds, M. P. (2011). The expansion of 'real estate tourism' in coastal areas: Its behavior and implications. *RASAALA: Recreation and Society in Africa, Asia, and Latin America, 2*(1), 51–70.

Benson, M. C. (2013). Postcoloniality and privilege in new lifestyle flows: The case of North Americans in Panama. *Mobilities, 8*(3), 313–330.

Bernier, E. T. (2003). El turismo residenciado y sus efectos en los destinos turísticos. *Estudios Turísticos, 155/156*, 45–70.

Casado-Diaz, M. A. (2016). Social capital in the sun: Bonding and bridging social capital among British retirees. In M. Benson & K. O'Reilly (Eds.), *Lifestyle migration: Expectations, aspirations and experiences* (pp. 87–102). New York City/Oxon: Routledge/Taylor & Francis Group.

Casado-Diaz, M., Kaiser, C., & Warnes, A. M. (2004). Northern European retired residents in nine southern European areas: Characteristics motivations and adjustment. *Ageing and Society, 24*(3), 353–381.

Covert, L. P. (2010). Defining a place, defining a nation: San Miguel de Allende through Mexican and foreign eyes. Dissertation for a Degree of Doctor of Philosophy within Yale University. Ann Arbor: Proquest.

Covert, L. P. (2017). *San Miguel de Allende: Mexicans, foreigners, and the making of a world heritage site*. Lincoln/London: University of Nebraska Press.

Croucher, S. L. (2011). *The other side of the fence: American migrants in Mexico*. Austin: University of Texas Press.

David, I., Eimermann, M., & Akerlund, U. (2015). An exploration of lifestyle mobility industry. In K. Torkington, I. David, & J. Sardinha (Eds.), *Practicing the good life: Lifestyle migration in practices* (pp. 138–116). Newcastle upon Tyne: Cambridge Scholars Publishing.

de Azuay, Gubernación. (2018, May 16). *Noticias. Gobernador del Azuay reconoció la labor voluntaria de los extranjeros en Cuenca*. https://www.gobazuay.gob.ec/gobernador-del-azuay-reconocio-la-labor-voluntaria-de-los-extranjeros-en-cuenca/. Accessed 19 June 2018.

Dixon, D., Murray, J., & Gelatt, J. (2006). *America's emigrants: U.S. retirement migration to Mexico and Panama*. Washington, DC: Migration Policy Institute.

Foulds, A. (2014). *Buying a colonial dream: The role of lifestyle migrants in the gentrification of the historic center of Granada, Nicaragua*. Theses and Dissertation – Geography, 18. https://uknowledge.uky.edu/geography_etds/18/. Lexington: University of Kentucky.

Grunenfelder-Elliker, B. (2011). Ir para volver – volver para retornar: Agrosubsistencia, y movilidad social bajo el impacto de la globalización en el Austro ecuatoriano. *Flasco Andes*.

Gustafson, P. (2016). Our home in Spain: Residential strategies in international retirement migration. In M. Benson & K. O'Reilly (Eds.), *Lifestyle migration: Expectations, aspirations and experiences* (pp. 69–86). New York City/Oxon: Routledge/Taylor & Francis Group.

Gustafson, P., & Laksfoss Cardozo, A. E. (2017). Language use and social inclusion in international retirement migration. *Social Inclusion, 5*(4), 69–77.

Haas, H. (2013). Volunteering in retirement migration: Meanings and functions of charitable activities for older British residents in Spain. *Ageing and Society, 33*, 1374–1400.

Hartshorne, J. K., Tenenbaum, J. B., & Pinker, S. (2018). A critical period for second language acquisition: Evidence from 2/3 million English speakers. *Cognition, 177*, 263–277.

Hayes, M. (2015). Introduction: The emerging lifestyle migration industry and geographies of transnationalism, mobility and displacement in Latin America. *Journal of Latin American Geography, 14*(1), 7–18.

Hayes, M. (2018). *Gringolandia: Lifestyle migration under late capitalism*. Minneapolis: University of Minnesota Press.

Hayes, M., & Carlson, J. (2018). Good guests and obnoxious gringos: Cosmopolitan ideals among North American migrants to Cuenca, Ecuador. *American Journal of Cultural Sociology, 6*(1), 189–211.

Herrera, G., & Martinez, L. P. (2015). ¿Tiempos de crisis, tiempos de retorno? Trayectorias migratorias, laborales y sociales de migrantes retornados en Ecuador. Estudios Políticos, (47), 221–241.

Huber, A., & O'Reilly, K. (2004). The construction of Heimat under conditions of individualised modernity: Swiss and British elderly migrants in Spain. *Ageing and Society, 24*(3), 327–351.

Huete, R., Mantecón, A., & Mazón, T. (2007). *La percepción de los impactos del turismo residencial por parte de la sociedad receptora. Comunicación presentada en las II Jornadas sobre turismo y sociedad*. Córdoba: IESA-CSIC. February 2008. http://rua.ua.es/dspace/handle/10045/14453. Accessed 20 Dec 2017.

Jokisch BD. (2014, November 24). *Ecuador: From mass emigration to return migration?*. Washington, DC: Migration Policy Institute. https://www.migrationpolicy.org/print/15140#. XLRgpuhKg2w. Accessed 15 April 2019.

Koutonin, M. R. (2015, March 13). Why are white people expats when the rest of us are immigrants? *The Guardian*. https://www.theguardian.com/global-development-professionals-network/2015/mar/13/white-people-expats-immigrants-migration?CMP=share_btn_fb. Accessed 17 June 2018.

López, E. (2017). *Estudio y Análisis para la creación de un "Programa de Capacitación en Idioma Inglés para Conductores de Taxi en la cuidad de Cuenca"*. Dissertation for a Degree from the Universidad del Azuay, Escuela de Estudios Internacionales. Cuenca: Universidad del Azuay.

Nudrali, O., & O'Reilly, K. (2016). Taking the risk: The British in Didim, Turkey. In M. Benson & K. O'Reilly (Eds.), *Lifestyle migration: Expectations, aspirations and experiences* (pp. 137–152). New York City/Oxon: Routledge/Taylor & Francis Group.

Rojas, V., LeBlanc, H. P., & Sunil, T. S. (2014). US retirement migration to Mexico: Understanding issues of adaptation, networking, and social integration. *International Migration and Integration, 15*(2), 257–273.

Serow, W. J. (2003). Economic consequences of retiree concentrations: A review of North American studies. *The Gerontologist, 43*(6), 897–903.

Terzian P. (2018, July 10). The world's top 15 cities. *Travel and Leisure*. https://www.travelandleisure.com/worlds-best/cities. Accessed 15 Apr 2019.

United Nations Educational, Scientific, and Cultural Organization (UNESCO). *World Heritage List*. https://whc.unesco.org/en/list/. Accessed 14 June 2019.

Valdivia, C. E. (2014). *En rol de las ONG en América Latina: Los desafíos de un presente cambiante*. Lima: Mesa de Articulación de Plataformas Nacionales y Redes Regionales de América Latina y el Caribe.

Williams, A. M., King, R., & Warnes, T. (1997). A place in the sun: International retirement migration from northern to southern Europe. *European Urban and Regional Studies, 4*(2), 115–134.

Wilson, P. (2013, September 24). Retiring in Latin America is easier than you may think. *Forbes*. https://www.forbes.com/sites/nextavenue/2013/09/24/retiring-in-latin-america-is-easier-than-you-may-think/#b9f76b115b3b. Accessed 20 Apr 2019.

# Chapter 5
# How Small Neighborhood Convenience Store Operators in Ecuador and Mexico View the Growing Presence of Retired Americans

Philip D. Sloane, Maria Gabriela Castro, and Erika Munshi

Neighborhood convenience stores ("tiendas de barrio") have an important role in the national and household economies in Latin America. Open daily, often for as many as 16 hours a day, tiendas vary in size but are typically quite small, often with one clerk located near the front of the store to monitor traffic and take payment from customers. Their traditional roles included serving a neighborhood, often selling small quantities of goods to persons who could not afford to buy in bulk, working in partnership with small businesses such as bread bakers or tortilla makers to market fresh products daily, providing credit to clients whose economic circumstances are day-to-day, and serving as neighborhood gathering places.

In contrast to neighborhood grocery stores in the United States and Canada, tiendas have been relatively resilient despite the introduction of supermarkets in their communities. This is part of a general trend of small business persistence in Latin America. It was estimated in 2009 that Latin America contained 69 million microbusinesses, which generated 70% of the region's employment and 20–22% of the gross national product (Añasco 2011). One reason for this trend appears to be the geographic centrality of tiendas, making them places that local inhabitants pass during their daily routines. In urban areas, it is not uncommon to find as many as a dozen small tiendas within walking distance of any given household (Pisani and Yoskowitz 2012).

In contrast to basic convenience stores found in the US, tiendas often have fresh produce and other food items. With their diverse product assortment, tiendas serve

P. D. Sloane (✉) · M. G. Castro · E. Munshi
University of North Carolina at Chapel Hill, Chapel Hill, NC, USA
e-mail: philip_sloane@med.unc.edu

© Springer Nature Switzerland AG 2020
P. D. Sloane et al. (eds.), *Retirement Migration from the U.S. to Latin American Colonial Cities*, International Perspectives on Aging 27,
https://doi.org/10.1007/978-3-030-33543-4_5

as accessible sources of household essentials. The most common products sold in Latin American neighborhood tiendas fall into four categories:

- staples for the kitchen, such as milk, bread, cooking oil, spices, pasta, cheese, and cereal;
- basic household products, such as soaps, sanitary pads and tampons, deodorants, oral care products, and hair products;
- alcohol and tobacco products; and
- snack foods, including candy, nuts, sodas, chocolates, cookies, crackers, and ice cream products.

In addition, depending on the tienda, there may be other specialty items, as well (Pisani and Yoskowitz 2012; Cámara de Proveedores y Canales de Distribución 2013).

Tiendas vary greatly in size, ranging from marginal operations to "dynamic enterprises"(Pisani and Yoskowitz 2012). Most are small, family-owned businesses, operated with the goal of sustaining the family that runs the store rather than as a profit-center (Mexperience 2018). As such, many tiendas run on a relatively thin margin. A study in Ecuador, for example, estimated that 35% of tiendas had sales volumes totaling less than $100 a day (Añasco 2011). In nearly 90% of tiendas, the owner-operator works full-time, with the more successful tiendas often headed by dynamic entrepreneurs, many of whom are women (Pisani and Yoskowitz 2012). Due to their relatively small size and family-oriented business operation, tiendas may have fewer regulatory requirements, a factor that can help maintain their viability by limiting the administrative burden on the owner-operators.

Still, there is concern that expansion of supermarkets in many metropolitan areas throughout Latin America, and to a lesser extent the growth of chain convenience stores such as OXXO, threaten the persistence, ubiquity, and social roles of the neighborhood tienda (Reardon and Berdegué 2002). Supermarkets, by their large size and aggregation into chains, have superior purchasing power, thereby wielding a dominant influence in the agrifood economy, often to the exclusion of small farms, processing, and distribution firms, and the small tiendas as well (Reardon and Berdegué 2002; Zamora 2005). Wholesale markets increasingly sell preferentially to large distribution centers, that almost exclusively then sell to supermarkets, which purchase in bulk and can negotiate lower per-item prices (Schwentesius and Gómez 2002). As a result, local tiendas have had to keep prices lower than would be ideal, and in some cases add products not available in chain convenience stores, such as fresh produce, freshly baked bread, and even fresh meat – activities that may improve or threaten their profitability (Mexperience 2018; Salazar 2016).

Supermarket and chain convenience store growth has not, however, led to the dominance that these enjoy in the United States and Canada. As of the first decade of the twenty-first century, according to one review, supermarkets held 90% of retail food sales in the United States, whereas in Mexico that figure was 45% (Traill 2006).

Still, the neighborhood tienda faces many challenges (Salazar 2016). Due to supermarket intrusion, tienda reliance is low in many areas, particularly in well-to-do neighborhoods, where automobile transport is more common and consumers have the resources to purchase items in bulk. In one study, for example, fewer than 5% of residents surveyed in high income neighborhoods in Mexico City relied on local tiendas for purchase of personal use items, whereas in low income neighborhoods the proportion ranged between 24% and 60%. Similar figures were noted regarding preferred purchase locations for daily household items (Duhau and Giglia 2007). Additional threats to survival of these microenterprises are neighborhood crime and insecurity; lack of business sophistication of the owner-operators; inability to obtain volume discounts from suppliers; lack of access to credit; and the increasing availability of other, better paying, work opportunities for women (Añasco 2011; Zamora 2005).

On the other hand, there are forces that to some extent are assisting neighborhood tiendas in remaining economically viable. Some relief has been provided by distribution companies that have specifically targeted local stores (Reardon and Berdegué 2002), and the development of microcredit options by banks and nongovernmental organizations (Añasco 2011). In addition, some products and distributors offer special discounts, exhibits, or promotions to small tienda owners, and some more entrepreneurial tienda operators stock niche items that appeal to particular customers (Salazar 2016).

Indeed, the continued flourishing of neighborhood tiendas in spite of supermarket growth has been surprising to many and is thought in part to represent differential values in Latin America compared to the United States, particularly in less affluent areas. In these neighborhoods, tiendas play a vital role in social dynamics, serving as "quasi-public spaces," supporting local information exchange, providing informal social support, and extending credit to regular customers in times of economic stress (Pisani and Yoskowitz 2012; Coen et al. 2008). These roles reflect community values that reinforce consumer use of tiendas – including the value of local meeting spaces, confidence engendered by personal relationships, neighborly cordiality, tradition, loyalty, gratitude, convenience, and miniaturization (purchasing in small amounts) (Ramirez Plazas 2008).

## 5.1 Selection of Tiendas for Research

As owners of small businesses, tienda shopkeepers are uniquely sensitive to shifts in prosperity and spending habits and to the personal circumstances of their clients. Since tiendas serve as public spaces where residents of all backgrounds shop, they offered a uniquely distilled viewpoint of the average resident's opinions towards North American retirees. Because of their relationship to the community and insight into the economy, tienda owners offered valuable information about the status of the economy as well as the perception of the impact of North American retirees on the economy.

Because of their ubiquity and role as social nexuses within local neighborhoods, our research study considered tienda operators to be an excellent, standardized population to survey as a means of gathering neighborhood perceptions of retirees from the U.S., Canada, and Western Europe, including their economic impact in the community. Tienda owners and employees also offered a contrast to the other local populations interviewed, who were largely middle and upper class, some of whom were in positions where they may not freely share their personal opinions on retiree presence.

The research team interviewed two groups of tienda operators based on the location of their business – a group from neighborhoods that were highly trafficked by retirees, and one from areas with little retiree presence. In addition to providing us with a range of interviewees, this method would allow us to compare and contrast tiendas in areas with either high or low retiree concentration. Figures 5.1 and 5.2 identify the general location of the tiendas where our interviews took place, with the location of the tiendas in each city placed on a density map of retiree concentration.

## 5.2 Measures and Data Collection

Research staff spent the first week in each city stratifying neighborhoods into either high or low-retiree concentration areas in order to sample six tiendas from each group. This geo-mapping was done with the assistance of the local, on-site coordinator within each city, with data provided by local interviewees to identify on a scale of 0 (none) to 3 (many) where retirees were concentrated on a map (see Figs. 5.1 and 5.2). Once patterns in neighborhoods became evident and were supported by additional field work, the research team toured each type of neighborhood in order to identify potential research participants. Through these means, research staff were able to reach the study's goal of interviewing 24 tienda operators total, six from each group per city, representing over a quarter of the local residents interviewed during the course of the project.

A standardized, open-ended interview in Spanish was administered to each participating tienda owner. It sought their opinions regarding immigrant retirees, including their experiences interacting with them, their views about the economy, and about governmental programs for retirees. The interview also included a series of Likert-scaled questions regarding the extent to which they agreed or disagreed with a variety of statements about retirees. These sections were administered to all groups interviewed during the course of the research. In addition to these sections, tienda operators were asked price information on five staple foods now and 5 years ago (milk, eggs, rice, sugar, and onions), and we conducted observations to identify from a checklist of fresh fruit and vegetables which were available in that tienda. We also took pictures, when given permission, of the entrance and the produce section. Details of research methods for this study are provided in the Appendix.

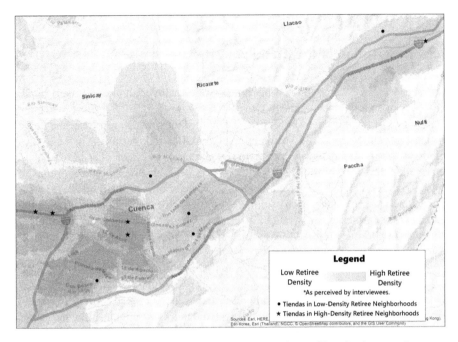

**Fig. 5.1** Location of tiendas studied in Cuenca superimposed on resident density map of concentration of retirees in the city. One tienda operator in each group declined to have their location identified

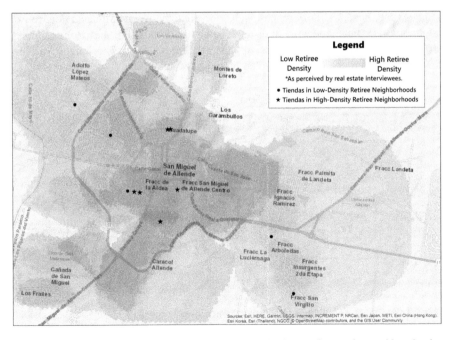

**Fig. 5.2** Location of tiendas studied in San Miguel de Allende superimposed on resident density map of concentration of retirees in the city

## 5.3   Description of the Tiendas Visited

The tiendas where we conducted interviews varied widely, though all were small, single-room operations with one or two clerks. Figures 5.3, 5.4, 5.5, and 5.6 display street views of the entrances to four tiendas in the study, which represent the range of tiendas seen. Many could be freely entered and exited, but a few had cages or windows that required the purchaser to ask the clerk to either open a gate or to retrieve the items for him or her. Tienda owners often lived above the store, as in Figs. 5.3 and 5.4. Snack foods and beverages were prominent in all tiendas visited, and all had kitchen staples such as milk, eggs, flour, rice, and onions. None had the abundance of produce one would expect at a supermarket, with the majority of stores having produce limited to non-refrigerated staples with long shelf lives, such as tomatoes, potatoes, and onions.

Variation between tiendas was quite common in both high-retiree (often wealthier) or low-retiree (often less wealthy) neighborhoods. While we had hypothesized that tiendas in high-retiree neighborhoods would have a greater variety of goods, particularly produce, our observations indicated that this was not a strong trend. We also looked within each tienda to see if certain vegetables and fruits were present, including both produce items that are grown in the two countries and those that are more commonly found in (and often imported from) the US or Canada. Figure 5.7 displays the proportion of tiendas that, at the time of our data collection visit, stocked selected fruits and vegetables. The most commonly available produce items, all of which were present in 80% or more of tiendas, included avocados, apples, oranges, and bananas. Differences were not great between high and low-retire neighborhoods, but tiendas in high-retiree neighborhoods did tend to have more variety, particularly of strawberries, apples, and limes – items that might particularly appeal to retirees from the United States and Canada.

**Figs. 5.3 and 5.4**  Photos of representative tiendas in non-retiree areas. 5.3 is of a tienda in middle class area of Cuenca with few or any foreign retires. 5.4 is of a tienda from a neighborhood on the outskirts of San Miguel de Allende that retirees rarely frequent

**Figs. 5.5 and 5.6** Photos from Cuenca demonstrating the wide range of tiendas in areas frequented by retirees. 5.5 is a relatively large, modern tienda with a sign in English and easy ingress and egress. 5.6 had a cage-like entrance – a feature that is typically found in poor, high-crime neighborhoods

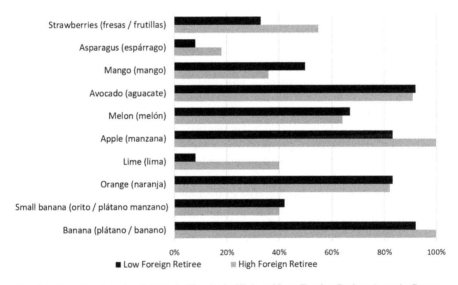

**Fig. 5.7** Fresh Produce Availability in Tiendas in High and Low Foreign Retiree Areas in Cuenca, Ecuador and San Miguel de Allende, Mexico

## 5.4  What Tienda Operators in Cuenca Said About the Status of the Economy

In general, tienda operators described Cuenca as a beautiful and well-tended city with reasonably good economic prospects to attract and host retirees, as well as to support tourism. However, they also alluded to socioeconomic disparities within the local population, considering this and the scarcity of jobs to be markers of a less robust state of the economy.

Tienda operators in Cuenca spoke in general terms about challenging economic conditions in Latin America, and the growth and decline of the economy in Ecuador in particular. They uniformly described an economic crisis affecting the region, with people struggling to find work and to support themselves. One commented, *"Currently the country is going through a severe economic crisis and there isn't enough work. There were too many demands on small industries, and many closed, creating unemployment."* Another considered that the difficulty of finding work incentivized people to go into business for themselves (such as opening a tienda), as this would provide a source of income. *"Right now, we're low on work. In general, there's a lot of competition and there just aren't any jobs. So, the majority of people start their own business in order to have income."*

Another common theme among the tienda operators was economic fluctuation. Interviewees frequently stated that economic prosperity was greater earlier in this century and attributed this to dollarization and increased government spending buoyed by high oil revenues. Recent years were given a more nuanced economic picture, in which rising prices were variously attributed to decreased agricultural production, immigrant retirees from the United States and other developed countries, and the influx of returning Ecuadorians (see Chap. 4 for a discussion of these immigrants). The economic change reported most often by our interviewees was increased prices for housing, goods, and services. As one tienda operator said, *"seven or eight years ago there was an economic boom ... then in the last four or five years it has diminished and almost disappeared ... now, because we are dollarized, prices keep rising and rising. In these last five, six years we have become one of the most expensive countries in Latin America."* Another stated that *"prices have gone up but, more importantly... prices are increasing by greater percentages than they ever have."*

From the tienda operators' perspectives, Ecuador's economy, while still tenuous, was in better shape than that of neighboring and nearby countries, driving migration from less economically stable areas in Colombia, Venezuela and Peru. Many interviewees reported concern that this type of migration was a security threat, due to bringing a "different type" of migrant that they perceived as a competitor for the already scarce local resources. This different type of migrant was also seen as requiring more security, as interviewees characterized them as uneducated and poor, therefore having a tendency to criminal behavior.

> *"In the last six months, many Venezuelans have migrated [to Ecuador].... They are not like the American that has his money, his retirement and lives here peacefully. Many migrate here because of economic problems. These are people who come without work, and that complicates matters....These are people without education ... That's a problem."*

In sum, a major theme from interviewees across different areas of Cuenca was of a regional economic crisis marked by unemployment superimposed on increasing costs. They agreed that this was driving unwanted immigration from neighboring countries, creating competition for jobs and possibly destabilizing local security. These ideas were equally reflected among high and low retiree area tienda operators.

## 5.5 What Tienda Operators in San Miguel de Allende Said About the Status of the Economy

The tienda operators we interviewed in San Miguel de Allende described a different economic picture from that described by tienda owners in Cuenca. They spoke of a strong global economy and increased international visibility for the city, that has created a more secure income stream for San Miguel. Specifically, they mentioned that having international consumers stabilized the local the economy by supporting it with foreign income. As one said: "*San Miguel has had more exposure at an international level. It's become more well-known, and we now have more national and international visitors. This has brought us a little more financial security.*" Thus, the boom in tourists and retirees was generally seen as a good thing by the tienda owners, although they, like interviewees in other professions, described significant downsides in terms of increased costs, gentrification, and reduction in local quality of life.

While tienda owners lauded the presence of immigrant retirees from other countries for their financial contributions, they also commonly spoke of inequity - both between the expat retirees and the local population and within the local population. The stronger economy benefited most people, but interviewees also thought that this economic growth amplified existing economic disparities within the population. They remarked that the increased cost and market for both labor and goods had created competition for local residents. They also expressed concerns about loss of purchasing power in the face of economic growth, though often expressing acceptance that this was the forward trajectory of a tourism-based economy.

> "*It's a touristic city and we are in the commercial center. Obviously life isn't cheap, but...we have this lifestyle. The cost of land has increased and many people from San Miguel can no longer afford it. But there's no other way.*"

Most tienda operators in San Miguel de Allende commented on the increased cost of labor and real estate and the increasing unaffordability of the city center for the local population. They noted that in the last 15 years, coincident with marked increases in tourist and international retirees, the costs of labor and real estate have increased rapidly in the city center. As a result, many local families have moved to areas less frequented by foreigners, where they could retain their spending power. One said "*Since San Miguel became a "magical city" ("pueblo mágico", a designation awarded to a number of cities by the national government, as part of a move to attract tourism),* (Government of Mexico 2019) *property values have increased dramatically. For example, depending on the area of San Miguel, the cost of basic food, the basic basket has increased. The city center is incredibly expensive. On the outskirts it's a bit more economical.*" Another said, "*Yes, I can buy a car, buy a house and I still have money. But I live in a less desirable area and at times it is insecure.*"

## 5.6   What Tienda Operators in Both Cities Said About the Impact of Immigrant Retirees from the United States and Other "Developed" Countries on the Economy and on Their Businesses

The tienda operators we interviewed in both cities were unanimous in agreeing that prices had risen. However, their opinions differed as to whether or not this was a result of retiree migration. Some considered that the willingness, or acquiescence, of expats to pay higher prices increased the cost of purchased goods for all. Others considered it natural for a growing economy to have inflation and increased costs, while lamenting that the minimum wages were not keeping up.

Tienda owners in both cities uniformly reported that immigrant retirees had a positive impact on the economy and that this benefit was distributed across the entire population for both small and large merchants as well as those that were recipients of the additional spending downstream of direct consumption by retirees.

> *"I'm a small merchant, but I get a lot of benefit. The woman out there on the street that sells fruit, she benefits as well, because they buy from her. Everyone benefits from immigration in the sense that there's a little more money overall."*

Descriptions of this economic impact included the following: (a) increased overall spending in areas such as tiendas, bars, restaurants, and taxis, as the retirees procured goods and services necessary to establish their semi-permanent households; (b) "incentivized spending" on items and services that are comparatively inexpensive for foreign expatriates, such as fresh produce and having a housekeeper; (c) new sources of income through demand for previously underutilized services, such as gardening, renovation, and dog walking; and (d) generation of employment that redistributed wealth and increased spending among the local population. One interviewer remarked:

> *"It's a cascade effect, because there are more people that can be gardeners, housekeepers, architects to build houses for those that are arriving, interpreters, and restaurant workers. The people who come here like to eat well. It's a cascade that can function well."*

Furthermore, retiree migration was viewed as not having some of the adverse consequences of other forms of immigration, in which foreigners arrive without money or seeking work, thereby adding to job competition or unemployment burden. Furthermore, the retirees were relatively complacent and therefore didn't appear to incur liabilities for the community. As one interviewee said: *"They spend their money here. It's a great economic support for us. And because they are elderly, there aren't problems with them, there's no security risk. It's not the type of immigration that makes you feel insecure."*

Most tienda operators we interviewed described retirees as a different type of customer, whose lower threshold to spend compounded the greater purchasing power of their income in the local economy. This reflects the fact that median retiree monthly incomes were considerably higher than the average monthly wage in both cities, thereby allowing them to afford "luxuries" such as paying inflated prices for foods and services.

Some tienda operators considered the direct benefits to be less on their own businesses than on such enterprises as bars and restaurants. A contrasting opinion, articulated by one interviewee in San Miguel, stated that bars and restaurants were frequented largely by short term tourists, and that the retirees more often patronized businesses such as tiendas. Reported benefits of retirees over tourists also involved providing employment for domestic workers such as cleaners, housekeepers and maintenance persons, thereby creating a sustained income for what would otherwise have been a seasonal operation.

*"Carpentry, metal work and gardening have gotten a lot more expensive. The need for gardening is a new development, because we haven't really taken to gardens here, nor to having someone who tends them."*

Some tienda operators in both cities expressed concern about the degree to which costs had increased due to an "Americanization" of the system. Negotiations with expat retirees introduced a shift from a previous flat rate for day labor to an hourly charge. This system of valorizing shorter time intervals allowed the laborer to determine their cost by project needs and the presence of a competing consumer allowed them to decline local customers unwilling to negotiate in this new fee structure. Interviewees felt that this competition doubly hurt local consumers – first, because they were competing with foreign spending power for bids on previously affordable work, and second, because the foreign consumers were willing to pay more and were less demanding in terms of work.

*"Almost no one wants to work in the homes of the Mexican population … because they have better pay and they are treated better [by foreigners]. We are more demanding; we ask for more physical labor. If I pay you, you need to clean everything, and even then I will go look for something else for you to clean. And [the foreigners] don't. They contract you to clean up this area, and that's good enough."*

Tienda operators we interviewed had differing opinions about the impact of retirees as employers on the job market. Many spoke of higher wages leading to competition for this high-cost employment, whereas others noted that some manual laborers felt underpaid by foreigners. In general, however, interviewees felt that the new phenomena of hourly (instead of daily) pay and payment in dollars (in Mexico) disincentivized workers from working for the local population and led many local workers to prefer working for expat retirees.

Some interviewees stated that foreigners "would pay any price for almost any item" that they wanted to purchase. For the expats, they felt, prices for many items, such as produce and kitchen staples, are inexpensive compared with what they would pay in the US or Europe; so, they often would not complain or negotiate with

**Table 5.1** Price of Typical Food Staples in Tiendas Surveyed and USA – in US Dollars; June/July 2018

|                          | San Miguel de Allende[b] | Cuenca        | U.S.A.[c] |
|--------------------------|--------------------------|---------------|-----------|
| A. Liter of Milk         | $0.86 (0.10)             | $0.91 (0.06)  | $0.85     |
| B. Dozen eggs            | $0.98 (0.26)             | $1.50 (0.35)  | $2.80     |
| C. Pound of Rice         | $0.57 (0.16)             | $0.67 (0.29)  | $0.71     |
| D. Pound of sugar        | $0.47 (0.05)             | $0.52 (0.19)  | $0.66     |
| E. Pound of onions       | $0.35 (0.10)             | $0.43 (0.12)  | $1.05     |
| **Bag of groceries**[a]  | **$3.24 (0.29)**         | **$3.95 (0.63)** | **$6.07** |

[a]A bag of groceries was defined as a liter of milk, a dozen eggs, a pound of rice, a pound of sugar, and a pound of onions
[b]Peso to dollar exchange rates calculated based on the exchange rate on July 11, 2018, at 1 peso to 0.052626 dollars
[c]Sources for US prices:
Milk: https://www.ams.usda.gov/sites/default/files/media/RetailMilkPrices2018.pdf. Based on conversion of price of a gallon of milk to a liter
Eggs: https://www.cookinglight.com/news/egg-prices-record-low
Rice: https://www.statista.com/statistics/236628/retail-price-of-white-rice-in-the-united-states/
Sugar: https://www.ers.usda.gov/data-products/sugar-and-sweeteners-yearbook-tables.aspx
Onions: https://www.ers.usda.gov/data-products/sugar-and-sweeteners-yearbook-tables.aspx

stated prices. Merchants also noted that retirees were more likely to be willing to buy, compared with locals who required some persuasion or who put in a bid for negotiation. Other respondents thought that over time the spending habits of the retirees conformed to those of the local population, for example, bargaining for lower prices.

Validation for these observations is provided by data collected by the study team on the cost of a "bag of common grocery items" at the study tiendas and comparison data from the United States. We defined a "bag of groceries" as a liter of milk, a dozen eggs, a pound of rice, a pound of sugar, and a pound of onions – staples that were present in all tiendas we visited. In San Miguel de Allende, the average price of these items in equivalent U.S. dollars was $3.24; in Cuenca it was $3.95. In contrast, the average price in the United States for a similar "bag of groceries" at the same month and year was $6.07 (Table 5.1).

## 5.7 Impact of Supermarkets and Chain Grocery Stores

A common theme of the tienda operators we interviewed was that their businesses were stressed economically. They acknowledged feeling pressure to save, having to be careful of their spending, and having to work hard to maintain their business. However, none mentioned that their business was in threat of closure or that they were planning to seek other employment. Several acknowledged that some business was lost to the mega-stores. However, there was only one mention of the mega-stores being a direct threat to a tienda operator's business viability; the others seemed to consider their business secure and described a co-existence, with tiendas serving a different purpose than mega-stores.

The majority of high retiree area store operators reported some impact of chain grocery stores, albeit minimal. For example, they reported that, despite going to the large chain stores for some purchases, the expat clientele would still come to the tiendas out of convenience. They also were aware that coming to the tiendas instead of a larger commercial store offered immediate access to goods, which is particularly important when one is unfamiliar with the area. One tienda operator put it this way: "*Even though there are big stores like Soriana, like Mega, like Comercial Mexicana, the first things they need, if they are in the area, they come to buy here. So they benefit me a lot.*"

## 5.8   Differences Between Responses of High vs Low Retiree Area Interviewees

The tienda operators in high retiree areas routinely acknowledged a positive impact of retirees on the local economy in their role as consumers. Among operators in the low-retiree areas, in contrast, responses were mixed, with many stating that the higher prices were not adequately offset by increases in business.

Comparison of data on a "bag of groceries" from tiendas in high and low retiree neighborhoods indicated that the price of a bag of groceries in the high retiree neighborhoods averaged 4.5% higher in San Miguel de Allende and 8.7% higher in Cuenca. Much variation was noted in this small sample, however, and so these differences were not statistically different (p = 0.4 in both cities).

## 5.9   Conclusion: Tienda Interviews as "Person in the Street" Perspectives on the Economic Impact of Retiree Immigration

Our study of tienda shop owners in Cuenca, Ecuador and San Miguel de Allende, Mexico yielded several insights about the status of the economy and small business resilience in colonial cities.

We found that tienda owners in Cuenca expressed more apprehension about the status of the economy in their city in comparison with tienda owners in San Miguel de Allende. While respondents in both cities agreed that there was an overall increase in the cost of living in their respective cities, tienda owners in Cuenca felt strongly that the increase in cost of living paired with regional economic challenges was driving instability both within Ecuador and between Ecuador and neighboring countries. In contrast, tienda owners in San Miguel de Allende expressed optimism that the increased cost of living was emblematic of San Miguel's transition into a thriving center for tourism and retirement.

Common to tienda owners in both cities was an acknowledgement that economic growth was not benefitting all demographics equally, and that it was at times augmenting existing disparities between socioeconomic groups. In San Miguel in particular, tienda owners commented that the transition to a stronger tourism industry made the housing market in the city center more competitive. As a result, local residents were choosing to move to the outskirts of the city where the properties were more affordable.

Regarding the impact of the North American retirees, we found that tienda owners in both cities reported that the retirees utilized their businesses, thus helping contribute to the resiliency of these small enterprises. While we hypothesized that the presence of North American retirees in Latin American neighborhoods would contribute to a trend toward supermarket dominance, most respondents reported continued viability as smaller grocery vendors despite supermarket presence and use of their services by retirees. It seems that, although the supermarket format is likely more familiar to the retirees, the proximity and accessibility of neighborhood tiendas, combined with their "authentic" Latin American nature, is compelling enough to earn their business. For the local population, tiendas continue to serve a different role than commercial supermarkets, offering a social space for people to meet and develop a relationship with their neighborhood tienda owner.

# References

Añasco, V. (2011). *Análisis situacional de las tiendas de barrio del sureste de quito: Un estudio de caso del ámbito de distribución de la empresa pronaca y la distribuidora Argel. Fondo bibliográfico de FLACSO Ecuador: Master's Thesis.* Quito, Eucador: FLASCO Sede Ecuador. https://repositorio.flacsoandes.edu.ec/handle/10469/5173. Accessed 26 Mar 2019.

Cámara de Proveedores y Canales de Distribución. (2013). *Bogotá: Boletín Retail, Number 41.* http://proyectos.andi.com.co/cpcd/Camara%20de%20Proveedores%20y%20Canales%20de%20Distribucin/Boletin%20Retail%20No%2041.pdf. Accessed 25 Mar 2019.

Coen, S. E., Ross, N. A., & Turner, S. (2008). "Without tiendas it's a dead neighborhood": The socio-economic importance of small trade stores in Cochabamba, Bolivia. *Cities, 25*(6), 327–339.

Duhau, E., & Giglia, A. (2007). Nuevas centralidades y prácticas de consumo en la Cuidad de México: del microcomercio al hipermercado. *Revista Eure, 33*(98), 77–95.

Government of Mexico. (2019). *Pueblos Mágicos.* https://www.pueblosmexico.com.mx/. Accessed 23 Apr 2019. Note: San Miguel de Allende participated in the program for many years but gave up this designation when it was awarded a UNESCO World Heritage designation.

Mexperience. (2018). *Just for your convenience.* https://www.mexperience.com/just-for-your-convenience/. Accessed 25 May 2019.

Pisani, M. J., & Yoskowitz, D. W. (2012). A study of small neighborhood tienditas in Central America. *Latin American Research Review, 47*(4), 116–138.

Ramirez Plazas, E. (2008). ¿Por qué las tiendas de barrio en Colombia no han fracasado a la llegada de las grandes cadenas de supermercados? *Etornos, 21*, 37–50.

Reardon, T., & Berdegué, J. A. (2002). The rapid rise of supermarkets in Latin America: Challenges and opportunities for development. *Development Policy Review, 20*(4), 371–388.

Salazar, F. (2016). La segmentación de las bodegas en América Latina. *Actualidad*. https://www.esan.edu.pe/conexion/actualidad/2016/05/27/la-segmentacion-de-las-bodegas-en-america-latina/. Accessed 23 Mar 2019.

Schwentesius, R., & Gómez, M. Á. (2002). Supermarkets in Mexico: Impacts on horticulture systems. *Development Policy Review, 20*(4), 487–502.

Traill, W. B. (2006). The rapid rise of supermarkets? *Development Policy Review, 24*(2), 163–174.

Zamora, M. (2005). *La rápida expansión de los supermercados en Ecuador y sus efectos en las cadenas agroalimentárias*. Quito: Ecuador Debate, 135–150.

# Chapter 6
# Obtaining Health Care as an International Retiree Living in Cuenca, Ecuador and San Miguel de Allende, Mexico

Philip D. Sloane, Johanna Silbersack, and Sheryl Zimmerman

Because of their age, retirees often have chronic illnesses, use health care services frequently, and consequently consider health care access a greater priority than do younger persons. Therefore, we anticipated that the availability of quality health care would be important to international retirees who are living abroad in Latin America.

There are many considerations related to receiving health care as an expatriate in Latin America; issues related to quality and health insurance rank highly among them.

- Quality of care. Latin American health systems vary widely. In both Mexico and Ecuador, there tend to be parallel public and private systems, with the majority of the population receiving care in relatively under-resourced public hospitals and clinics while the minority pay out of pocket for private care. Wide disparities in care quality and access are reported between the two systems, and private care is praised as much less costly and far more personal than care in the United States (Pan American Health Organization 2017). However, for very complex problems, the public care system, with its academic hospitals, may have better resources.
- Health insurance coverage in America. Unlike the European Union, which has health services reciprocity between countries, Medicare – the government health insurance program for Americans who are old or disabled – provides no coverage outside of the United States. A few health insurance programs do provide such coverage – some state teachers retirement programs, for example – but they are quite rare. As a result, retirees in Mexico and Ecuador have three options: go "home" to obtain health care, enroll in an in-country health insurance program, or pay out of pocket.

P. D. Sloane (✉) · J. Silbersack · S. Zimmerman
University of North Carolina at Chapel Hill, Chapel Hill, NC, USA
e-mail: philip_sloane@med.unc.edu

© Springer Nature Switzerland AG 2020
P. D. Sloane et al. (eds.), *Retirement Migration from the U.S. to Latin American Colonial Cities*, International Perspectives on Aging 27,
https://doi.org/10.1007/978-3-030-33543-4_6

- Availability and quality of health insurance coverage in the host country. Retirees identify two general sources for in-country health insurance: (a) government health care for seniors, which is available to expatriate retirees after a waiting period and provides access to public clinics and hospitals for a monthly fee; and (b) a variety of private health plans, ranging from packages provided by clinics for outpatient services only to more comprehensive (and therefore more expensive) plans.

Many of the issues regarding health care for expatriate retirees in Latin America were illuminated by an earlier study, in which our research team interviewed 46 retirees from the United States who were living in Mexico or Panama and had been hospitalized (61%) or had a chronic health condition (78%) requiring ongoing medical care. Respondents largely used the private health care system, and compared to their experiences in the U.S., they praised their doctors for taking more time, charging far less, and going out of their way to provide assistance – for example, doctors providing their cell phone number or personally driving a patient to the hospital. Dental care was also generally praised for its high quality and low cost. Emergency services, on the other hand, were reported to be much less available and of lower quality than in the states, and hospital care was regarded as uneven. The biggest issue, however, was paying for hospital care. One uninsured interviewee, for example, had been hit by a motor vehicle while crossing a street, asked to be transported to a private hospital, stayed for several months of treatment and rehabilitation, and found that the charges largely exhausted her savings; another reported spending a large sum to return to the U.S. by private jet for emergency surgery (Sloane et al. 2013).

Given this background, we were interested in knowing how retirees from the United States, Canada, and western Europe who were living in Cuenca, Ecuador and San Miguel de Allende, Mexico accessed health care, and what their views were of the health care options in those cities. We also were interested in what representatives of the local population thought of the expatriate retirees as users of their public health systems.

## 6.1 Overview of the Health Care Systems in Mexico and Ecuador

### 6.1.1 Mexico

Mexico's modern health care system dates back to the 1940s, with the creation of the Secretariat de Salud and the Instituto Mexicano del Seguro Social (IMSS). The Secretariat de Salud was established as the government arm to oversee health care and the IMSS. Both of these organizations continue to be integral to Mexico's health care system, and in fact, both hold virtually the same roles now as they had in 1943 (Castro 2014; OECD 2016). Some substantial changes have occurred of course;

over the decades, the system has evolved to expand coverage and accessibility. A major turning point occurred in 1983, with an amendment to the Mexican Constitution (Castro 2014; OECD 2016) that stated that basic health care is a right of all Mexicans (Alcalde-Rabanal et al. 2017).

Currently, the Mexican health care system is divided into two primary segments: the private system and the public system (Pan American Health Organization 2017). The private system is a network of hospitals, clinics, and providers accessed either by paying into a private health insurance company or by paying out-of-pocket for services. However, this system is inaccessible for most Mexicans, due to its high prices. In fact, private health insurance covers only 7% of the Mexican population, although many individuals choose to use private services by paying out-of-pocket when affordable (OECD 2016).

The majority of Mexicans are served by the public sector, which can be broken into two types: social security institutions that provide services to Mexicans in the formal employment sector, and institutions that provide services to non-employed, or informally-employed, individuals (Pan American Health Organization 2017; Castro 2014). The social security arm of the public system is composed of a number of organizations: the IMSS, the Instituo de Seguridad y Servicios Sociales para los Trabajadores del Estado (ISSSTE) for government employees, and a number of others that separately service Army, Navy, petroleum, and other municipality workers (Pan American Health Organization 2017; Castro 2014; OECD 2016). These social security systems are paid for through government, employer, and worker contributions (Alcalde-Rabanal et al. 2017).

The second public sector health care system, offered through the Sistema de Protección Social en Salud (SPSS) and Seguro Popular, was launched in 2003 with the intention of offering health care to all uninsured individuals (Castro 2014; Alcalde-Rabanal et al. 2017). By 2011, 49 million individuals were covered by Seguro Popular (Castro 2014), and it continued to expand coverage, to more than 57 million individuals, by 2014 (OECD 2016). Even with these impressive numbers of individuals covered, up to 18% of the population continued to experience difficulties accessing health care in 2014 (Pan American Health Organization 2017).

Unlike in the United States, each of these health care systems works independently. Each sector, private and public, owns its own clinics and hospitals, and has its own affiliated doctors. This independence essentially means that if a patient is being cared for in a facility that does not correspond to their insurance, he or she would be transferred to the correct hospital once stable (OECD 2016).

In response to concerns about quality and accessibility of health care, Mexican President Andrés Manuel López Obrador announced that he was implementing a plan to eliminate the Seguro Popular (Morales et al. 2019). Part of President Obrador's platform when he ran for president and was elected in 2018 was advocacy of universal health care as a right guaranteed by the federal government. Institution of the new changes, which will be rolled out over several years, is to eliminate a patchwork of insurance and health care systems within the states and to create a national health care system (Government of Mexico 2019). As of the writing of this chapter (July, 2019), its impact for international retirees is unclear.

## 6.1.2 Ecuador

Ecuador has recently put into place health care reforms that parallel Mexico's. In 2008, a new Constitution established that free basic, public healthcare was a right of all Ecuadorian citizens and was to be regulated and led by a National Health System. A number of Articles of the Constitution give further specifications, including free health care, specialized care, and medications to various population groups, including elderly persons (Pan American Health Organization 2017; National Assembly of Ecuador 2008). This constitutional reform was a major overhaul from Ecuador's previous system, which had been plagued by chronically low spending due to constantly shifting governments (Aldulaimi and Mora 2017).

Ecuador's health system has been split into two main sectors, private and public. The private system is primarily used by the upper-middle class population of Ecuadorians; in fact, due to its expense, the private system covers only 3% of Ecuadorians (Aldulaimi and Mora 2017). It is composed of for-profit hospitals, clinics and companies (Lucio et al. 2011), and is accessible through private health insurance companies, prepaid plans from medical providers (Pan American Health Organization 2017), and out-of-pocket payments for specific services or visits (Lucio et al. 2011). In Cuenca, retirees tend toward the private options, with almost 61% of immigrants opting for private providers (Álvarez et al. 2017), despite the higher costs.

Similarly to Mexico's, the public system is composed of two overarching arms, one geared toward the uninsured population and one that offers social security services to the working population (Lucio et al. 2011). The Ministry of Public Health (MSP) and the Ministry of Economic and Social Inclusion (MIES), which offer health care free of charge to uninsured, unsalaried Ecuadorians, covered approximately 51% of the population in 2011 (Lucio et al. 2011). This care is drawn from public funds, allocated to the Ministry of Public Health by the Ministry of Finance (Villacres and Mena 2017).

The social security healthcare system is dominated by the Instituto Ecuatoriano de Seguridad Social (IESS), although like Mexico there are specific subsystems for sectors of workers, including the army and the national police force (Villacres and Mena 2017). The IESS, the largest of the social security organizations, covers approximately 20% of Ecuadorians (Lucio et al. 2011). The funds for each of these organizations come from state, employer, and employee contributions. It is also possible to voluntarily contribute and join the IESS, even without an employer. Through this mechanism, 44% of the immigrant retirees in Cuenca have chosen the IESS, although only 27% actively use it (Álvarez et al. 2017).

Following the health care reformation, Ecuador has made strides in accessibility and use; for example, health care spending increased from 1.15 billion in 2010 to 2.57 billion in 2015 (Pan American Health Organization 2017), doctor visits doubled between 2006 and 2012 (Aldulaimi and Mora 2017), and Ecuador has been ranked high on health care ranking lists such as Bloomberg Health Care Efficiency – in fact, above the United States in 2018 (Miller and Lu 2018).

## 6.2 Methods

Data reported in this chapter were gathered as part of a multi-methods study of the impact of retirees from the United States, Canada, and Western Europe who had moved to and were living in the historic colonial cities of Cuenca, Ecuador and San Miguel de Allende, Mexico. In addition to academic and internet sources cited in this chapter, original data were gathered from (a) an online survey of retirees in the two study cities, (b) interviews (in Spanish) with samples of local (i.e., native-born) residents living in the study cities, and (c) interviews with retirees who were actively using health care services. Appendix A provides details about the study methods; the three data sources are described in brief below.

- Online surveys were conducted June–August, 2018 in the two study cities. A separate survey was created for each city, but the questions were identical. Questions analyzed for this chapter include questions about availability and value of access to the government health care system after a waiting period; the extent to which respondents agree or disagree with the statement "I live here because the health care is affordable;" and (on a scale from 0 to 100) the percentage of their health care that they obtained in the country (i.e., in either Ecuador or Mexico).
- All of the 79 local residents interviewed in the two study cities were asked the following open-ended question about their use of the government health care system: "One of the benefits that immigrant retirees often want is access to local health care. What is your opinion about whether immigrant retirees should be able to enroll in the government health services of [name of country], and in what circumstances should that occur?" Of the interviewees, six in each city worked in health care; they were a purposive sample that include hospital administrative personnel, physicians, dentists, and staff of long-term care settings or home care programs. The entire content of all interviews was coded using NVivo software and analyzed as described in the Appendix. The following codes were analyzed for this chapter: (a) Health Care - current state, (b) Health Care - attitude toward retiree use, (c) Health Care – quality, and (d) Health Care – cost.
- During data collection visits to the study cities, our research team networked with local retirees and recorded relevant information about use of and opinions regarding health care obtained. We also inquired about and sought to interview in each city at least 3 persons aged 65 and older who were using formal long-term care services; results of those interviews are largely presented in Chap. 7; however, comments relating to health care services in general are included in this chapter.

## 6.3 Retiree Activities, Experiences, and Attitudes Regarding Health Care

Survey responses were received from 424 retirees in Cuenca and 325 in San Miguel de Allende. Very few reported having relocated for health care – less than 2% in Cuenca and fewer than 1% in San Miguel de Allende identified "affordable health care" as their primary motivation for moving (Table 6.1). This is despite the fact that

**Table 6.1** Internet survey responses of immigrant retirees (aged 55+) in Cuenca, Ecuador and San Miguel de Allende, Mexico regarding perceptions of health care in those cities

| Survey item and responses | | Percent responding, by city | | P-value for difference between cities |
|---|---|---|---|---|
| | | Cuenca (N = 424) | San Miguel de Allende (n = 325) | |
| Primary motivation for moving to City | Affordable health care | 1.9% | 0.6% | 0.13[a] |
| | Any other response | 94.6% | 95.7% | |
| | No response | 3.5% | 3.7% | |
| Perception that government health care system is available after a waiting period? | Yes | 80.0% | 62.8% | <0.001[a] |
| | No | 1.6% | 7.1% | |
| | Don't know | 5.0% | 14.8% | |
| | Did not answer | 13.4% | 15.4% | |
| Attitude regarding usefulness of access to government health care (if available) | Much benefit | 45.8% | 21.8% | <0.001[b] |
| | Some benefit | 19.8% | 24.6% | |
| | Little/no benefit | 13.7% | 25.5% | |
| | Did not answer | 20.8% | 28.0% | |
| Response to the survey item "I live here because the health care is affordable" | | N = 379 | N = 288 | <0.001[c] |
| | Strongly agree | 16.1% | 6.6% | |
| | Agree | 22.4% | 17.7% | |
| | Somewhat agree | 26.4% | 25.4% | |
| | Neither agree nor disagree | 15.3% | 23.3% | |
| | Somewhat disagree | 5.0% | 4.2% | |
| | Disagree | 9.0% | 13.2% | |
| | Strongly disagree | 5.8% | 9.7% | |

[a]Analyzed using Chi-Square statistic
[b]Analyzed using the Cochran–Mantel–Haenszel test
[c]Analysis using a student's t-test or the Wilcoxon two-sample Mann-Whitney test yielded similar results

a sizeable proportion of respondents (27% in Cuenca and 22% in San Miguel de Allende) were aged 55–64 – a group that typically does not qualify for Medicare in the United States and that therefore would be expected to value less costly health care. Most likely the relative health of those who relocate in retirement explains these findings (La Parra and Mateo 2008).

Despite not having relocated because of the health care resources, about two-thirds of respondents (65%) in Cuenca responded affirmatively to the statement "I live here because the health care is affordable." That figure in San Miguel de Allende was lower (50%), with the difference being statistically significant at p < 0.001 (Table 6.1).

As noted earlier, both Mexico and Ecuador have public health care systems that allow expat retirees to join for a fee after a waiting period. Among the expat retiree community, opinions about the availability and value of this benefit varied, with public health care in Ecuador being generally felt to be both more accessible and of greater benefit than that in Mexico. Indeed, respondents to our online survey in Cuenca were more likely to be aware of the availability of access to the government health care system than were those in San Miguel de Allende (80% vs 63%, p < 0.001), and over twice as many in Ecuador than Mexico considered this access to be of "much benefit" to themselves (46% vs 22%, p < 0.001). These results are displayed in Table 6.1.

Use of health care resources by the immigrant retirees paralleled these attitudes toward the local health care. Over two-thirds of residents of Cuenca reported obtaining 100% of their health care in country, while the corresponding proportion is San Miguel de Allende was 43% (Fig. 6.1).

In comments on the surveys, in interviews, and in informal conversation, the expat retirees had a great deal to say about health care. Much revolved around how

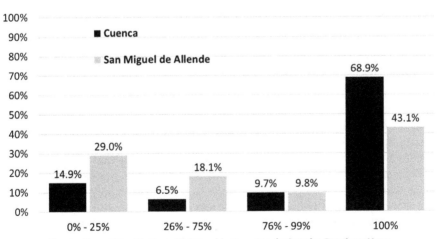

**Fig. 6.1** Responses of immigrant retirees in Cuenca (N = 370) and San Miguel de Allende (N = 276) regarding the proportion of their health care they received in-country

to obtain quality care, differences between Latin American health care and U.S. health care (since the vast majority were from the United States), and about how to cobble together resources to obtain health care coverage.

One common theme involved how to find a good doctor and obtain high quality primary care in the private system. Distrust of the public system in both countries was common. One typical story explained how an expat, wanting to obtain the least expensive care, had gone to a public doctor, where he received, *"the wrong dosage and medication."* As a result, many expats use private, fee-for service care, which in both cities was considered very good and quite reasonably priced. Private doctors were typically reported to spend more time and appear less hurried than their counterparts in the U.S., often being described as less test-oriented, instead spending more time taking a history and providing instructions on self-management. Doctors routinely give their cell phone numbers to patients. As one expat said, *"the culture of care is less concerned about liability and more concerned about the person."* Several expats talked about greater unevenness in quality compared to the U.S., however, and of the need to network with other expats to find a "good" doctor – one of whose characteristics was good English skills.

Distrust of U.S. health care and stories of bad care "up north" (in the U.S.) was another theme. "Too many medicines," "too many procedures," "too expensive", and "unsafe" were commonly expressed opinions. One said, *"I think going there for care is silly. All the people who do that, go up to the hospital there and die....It's incredible. They all die. They go, 'Oh I'm going to Houston, I'm going to have a blood clot removed in Houston.' Bingo. Dead."* Another told of a man who had chest pain in the U.S., was told he had a hiatal hernia, and was later seen in Cuenca, given a treadmill test, and diagnosed with severe heart disease. *"He went to Quito and had a triple bypass."*

Another common theme was how much less expensive care was in both countries than in the United States. Physician office visit costs were reported to be much lower than in the states ($45–$50 per visit was a common fee). As one interviewee said, *"It's so cheap you wonder what the hell is wrong with America. Because it's better care than in America by far."*

Still, paying for health care is a challenge for Americans living abroad. The big-ticket item, as elsewhere around the world, is hospital care. In the story presented above about the man who went to Quito for a triple bypass, the final bill was $50,000. Although this was about half as much as such a hospitalization would have cost in the United States (Gokhale 2013), the man had no health insurance and therefore had to pay out of pocket – a common situation because Medicare only pays for care provided within the United States. Furthermore, post-hospital arrangements and care are entirely up to the family – a problem especially for retirees who live alone and have a limited support system. As one expat said, *"When you get discharged from the hospital....there is no info about post-discharge care; so you have to be your own advocate."*

The high cost of hospitalization leads many immigrant retirees to look for arrangements that will defray those expenses in case of serious illness. Available options include enrolling in the public health care system; buying a catastrophic health insurance policy; and purchasing evacuation insurance that will allow return to the U.S. Each has advantages, disadvantages, and costs:

- The public health care systems do not have restrictions for pre-existing conditions, deductibles, or expenditure limits, while private insurance programs do. However, public hospitals are often described as of questionable quality, with long wait times, limited resources, and less technical know-how. This point is especially true in Mexico, whereas more expats in Cuenca praised the general quality of the public hospital system there.
- Catastrophic health insurance is quite expensive, and retirees expressed concerns about deductibles and other barriers to receiving reimbursement for hospital costs.
- Evacuation insurance allows persons who are abroad to return to the United States in case of illness, so that they can take advantage of Medicare and the U.S. health care system. Such a plan will work well for subacute or elective procedures, such as surgery for chronic back pain, gallstones, or a kidney stone. More emergent problems, however, such as an auto accident with multiple injuries, or a sudden, severe heart attack, would likely require in-country stabilization and care.

The majority of retirees tend to cobble together a combination of private health insurance, participation in the country's public health plan, out-of-pocket payments, and trips back to the U.S. to obtain health care services under Medicare (Sloane et al. 2013; Andrews 2019). In the process, they must be resourceful, do research, and serve as their own advocates. Assistance with these processes is most commonly provided by retiree blogs and other informal communication networks. In addition, some individuals have developed businesses to advise retirees on health insurance options, and provide educational workshops on emergency preparedness.

A perhaps remarkable finding from our interviews was how few retired Americans talked about the fact that Medicare did not provide coverage. Only a few bemoaned the "unfairness" of having paid into the Medicare program during their working years and then being denied coverage because they were living abroad; instead, most seemed to accept the fact that this was so, a feeling that was eased by cynicism about American health care in general and by a realization that expanding Medicare to Latin America would face insurmountable political, legal, and regulatory challenges (Bustamante et al. 2012).

Obtaining access to the public health insurance programs in both countries is a feasible option, but enrolling in both systems requires multiple administrative, eligibility, and payment processes. In our interviews, we heard many stories about the challenges of the process and about creative, at times underhanded, approaches used by expats to gain access to the system. One expat interviewee in Mexico, for example, mentioned that there were many "schemes" to obtain government health insurance, including bribing officials to bypass the "line" or to waive required paperwork, or paying a bilingual "facilitator" for help with the process. In Ecuador, the public health insurance program is theoretically available to all, including immigrants, due to a constitutional principle that all citizens have the right to voluntarily enroll in a system of universal health care. Foreigners must meet a residency requirement but can enroll earlier by registering as owners of a small business, which some retirees have creatively done by declaring as a business such activities as making cookies or providing massage services. Enrollment requires a fee based on one's declared income, which, according to one Ecuadorian hospital official, averages around $200 a month. However, our interviews with retirees indicated that many fail to declare all of their income, to pay less for health insurance.

Not all retirees have a comprehensive health care plan. For many, figuring out the intricacies of obtaining and paying for U.S.-style health services is a low priority. Most people we encountered were health conscious, exercised regularly, and were relatively young compared to the average age of retirees. Furthermore, by virtue of relocating internationally, they might be considered risk takers, and in that context it is not surprising that many chose not to make comprehensive health insurance a top priority. In addition, as noted earlier, many distrusted the U.S. health system and what they perceived as an unhealthy high-drug, high-tech approach to health care, and its tendency to keep people alive far beyond the point at which, in their estimation, life is no longer worth living. As one said: *"I want to live in a culture that lets you die. I think it's very important to be able to die in a natural way. And I hated what they did for my father. He couldn't talk anymore but you could see in his eyes that he wanted them to turn everything off and they wouldn't do it."*

Indeed, another theme that arose in discussions with expats, and that was reinforced by interviews with local residents in the two cities, is that retirees do die and that some of them die alone without social support. In Cuenca, three separate expat interviews on the same day related anecdotes about older persons who had died alone. One told of an older American who died alone in an apartment, and the body was not found for 2 weeks; another told of an older American who was hit by a car, died, and no one knew who the person was; a third mentioned that a body was currently in the morgue and no one could identify it. Similar stories were reported in San Miguel de Allende, including one that can be found online in a media report (Bickel 2017). Indeed, an often-unappreciated role of the U.S. consulate in both cities is to help identify deceased Americans and arrange a disposition for Americans who have been hospitalized, require post-discharge care, and have no family or friends to help with the arrangements.

## 6.4   Local Resident Attitudes and Comments About Immigrant Retiree Health Care

Our interviews with local residents in both cities tended to reinforce what we had heard from the immigrant retirees. Common themes included that the retirees often prefer private doctors over the public health care system; that they tend to gravitate toward doctors and nurses who speaks English; and that certain medical professionals in each city become known for attending to the retirees. Still, there were differences compared to the retiree survey results, which in some cases were expanded upon by our local resident interviews.

### 6.4.1 Attitudes and Opinions of Local Interviewees in Cuenca

Local residents we interviewed were aware that health care was relatively inexpensive and believed it to be of high quality. Some acknowledged that the system has drawbacks, such as long waiting times and resource limitations. Still, a theme among multiple interviewees was that the high quality and relatively low cost of public health care in Ecuador was attractive to international migrants. *"(A man) came from Florida,"* the interviewee said. *"A dental treatment there would cost $5,000, $6,000 dollars. Here even the worst cases won't cost as much as a thousand."* Several noted, however, that hospital procedures could be quite costly and could be "catastrophic" for an individual who lacked insurance. One noted, as an example, that placing a cardiac stent would cost between $15,000 and $30,000.

In that context, a pervasive theme among the local residents we interviewed was that everyone in the country was guaranteed access to health care by their constitution, and that this right extended to immigrants as well as citizens. A number of interviewees took obvious pride in their country's dedication to providing health care to all. *"They have a right, just like us,"* was a typical response to questions about immigrant retiree use of the system. Another gave this example:

> *"We have a public enterprise that carries all medicines, called Farmasol. The medicines are extremely inexpensive, at a very good price. Retirees have access to special discounts with a card. If you are a North American retiree you can obtain them, but a Cuencan retiree can as well. So, the idea is that all the benefits exist for all the community – foreigners as well as nationals."*

A significant minority, however, voiced resentment that the retirees were able to use the public health care system and pay what they perceived to be a small monthly enrollment fee. Several mentioned that the retirees, being older, use many medical services, and worried that this use could have a negative economic impact on the country's health care system. Some interviewees voiced a desire to see the retirees pay more and, in some cases, not use the public health care system at all. Considering the stories of some retirees under-declaring their income to pay less, these comments are not necessarily surprising. One said, *"The first retirees who came to Cuenca paid their medical bills out of their own wallet. Later they learned that using the public system was less expensive. They [now] use services that are for our own citizens."* Another said, *"I have contributed for a long time"* and went on to suggest that the retirees, because they had not done so, should purchase their own health insurance. *"I think that the public resources are for those who need them,"* another said, *"and that the retirees can pay for private insurance, which doesn't cost a lot."* A third interviewee voiced a similar opinion, saying that by paying a modest amount to gain access to public services, the immigrant retirees *"occupy space that poor Ecuadorians need."*

## 6.4.2  Attitudes and Opinions of Local Interviewees in San Miguel de Allende

As in Cuenca, local interviewees in San Miguel de Allende concurred that health care was less expensive in Mexico than in the U.S. There was, however, less consensus about the quality of the public health care system than we heard from natives in Cuenca, and also confirmation that the retirees tended to use the private system less and go elsewhere more for health care, as was demonstrated by our retiree survey (Fig. 6.1).

A major theme was that the public health care system had many weaknesses. Interviewees talked about a lack of resources, long waiting times, and inferior care compared with the private system. One native interviewee explained it as follows: *"The public health care system in Mexico is very bad. In fact, legally, all workers must have access to health care – medical services, hospitalization, all. But, because the number of people working for businesses that provide health care has dropped and Mexicans are living longer, now the public medical services in Mexico have very big financial problems."*

In part because of the inadequacy of the public system, San Miguel de Allende was seen as having a robust, relatively high-quality private system, which retirees preferred. Interviewees talked about bilingual general medical doctors, some of whom were trained "up north," and a new private hospital. Dental care was also described as excellent and inexpensive, with many bilingual dentists. One interviewee explained: *"They prefer private medical services because unfortunately the public medical services at times are lacking in resources. Thus, for them it is more practical....at least they have the ability to pay better than the people from here in Mexico, the lower classes, who have to use the public services."*

Few objected to the retirees enrolling in the public health insurance system, however, which was seen, similarly to in Ecuador, as a right. They also acknowledged that many retirees did not enroll, and that if they did, it was largely as a back-up in case of a major, costly care need. Typical comments included *"It is a right and it's good that we support it"*, *"everyone has the right to be attended to when sick"*, and *"yes, because it is a federal program and open to all, but only for residents (not tourists)."* Another acknowledged that some of the retirees have limited resources: *"Some come only with their pension,"* the interviewee said, *"and their pension sometimes is minimal for how expensive San Miguel is.....And so if they enroll (in the public system), I think it's fine."* One respondent had an additional reason for favoring the idea of immigrant retirees enrolling in the public healthcare system – *"It would raise the quality"*.

There was again, however, a significant minority who felt that the retirees should not use the public system. *"I believe that the national health services are saturated; thus, I don't know how we can give attention to the foreigners,"* one said. One additional reason, expressed by a few interviewees, is that some Americans demanded preferential treatment to avoid long waits in public clinics. One said, *"disgracefully they insist on being seen before others in the hospital – that is, to leap over people who were already there."*

To meet the private health care needs of the growing retiree population, over the past few years several new services have been developed in San Miguel. They include a new private hospital in the city; establishment of a private emergency care and ambulance service in town (Ambulancias Privadas SEETS); and creation of a technical training program in personal care of older persons, similar to nursing assistant training programs in the United States, supported by a collaboration between the city government and the state of Guanajato. The city has also seen an influx of health professionals from larger cities in Mexico, such as Querétaro and Guadalajara, including English-speaking physicians.

For major health needs, such as a cardiac procedure or a joint replacement, retirees frequently seek care outside of San Miguel de Allende, generally in Querétaro, an hour away, which has larger hospitals and a greater variety of specialty services; alternately, they return to their country of origin. *"Many go to Querétaro and are feeling that the medical attention in Mexico is better than in the United States,"* one interviewee said, *"but in the end they have many fears, and they go to the United States. However, they are fearful for various cultural reasons, and, really, medical attention in Mexico is good enough."* This statement reflects a perception that the retirees are conflicted in their use of these services. They desire to pay less than the highly expensive alternatives in the U.S., but they tend to not trust local health care, particularly in a setting where care is largely delivered in a language that they do not completely understand.

## 6.5   Conclusions

Obtaining, coordinating, and paying for health care is a major task and significant challenge for older people everywhere. Thus, it is no surprise that obtaining health care is a complex and often challenging issue for retirees who have chosen to relocate internationally to Cuenca, Ecuador and San Miguel de Allende, Mexico. The survey and interview results reported in this chapter provide a complex picture of varied opinions and experiences, with a number of themes that are similar across the two cities and countries, as well as some differences in the attitudes and approaches in the two locations. In both settings, health care is generally much less expensive than in the United States, but even so, out-of-pocket costs for hospitalizations and major procedures generally well exceed the average retiree's resources. As a result, retirees tend to create a personal patchwork of such things as private insurance programs, enrollment in public insurance, out-of-pocket payments, plans to return home for high-cost care, and wishful thinking that good health will persist for the foreseeable future. The balance among different sources of care varies across the two cities, with retirees in Cuenca using local resources more frequently (including the public system), and retirees in San Miguel de Allende more often going to another city or back to their country of origin for costly care, such as medical procedures.

On the whole, local residents' impressions of retirees using health care in their cities were favorable; many took pride in the fact that their governments considered health care a right rather than a privilege. However, some concerns were raised. These included retirees' lack of a history of having paid into the systems, as well as the impression that, because the majority of retirees can afford the private-pay system, they should not over-burden the public systems.

It is important to continue tracking retirees' use of the local health care systems and insurance as the two retirement destinations continue to mature. For example, as San Miguel de Allende has experienced higher number of retirees, there have been changes to their health care landscape including new hospital, emergency care and ambulance service. It remains to be seen whether similar processes will occur in Cuenca. Furthermore, the effects of recent initiatives by President Obrador in Mexico to overhaul the system are yet to be seen.

In sum, health care is a dynamic issue, and health care systems in both countries are continuously evolving, as are the health and care needs of each individual retiree.

# References

Alcalde-Rabanal, J., Becerril-Montekio, V., Montañez-Hernandez, J., Espinosa-Henao, O., Lozano, R., García-Bello, L., Lagunas-Alarcon, E., & Torres-Grimaldo, A. (2017). *Primary Health Care Systems (PRIMASYS): Case study from Mexico.* Geneva: World Health Organization. License: CC BY-NC-SA 3.0 IGO.

Aldulaimi, S., & Mora, F. E. (2017). A primary care system to improve health care efficiency: Lessons from Ecuador. *Journal of American Board of Family Medicine, 30*(3), 380–383.

Álvarez, M. G., Guerrero, P. O., & Herrera, L. P. (2017). *Estudio sobre los impactos socio-económicos en Cuenca de la migración residencial de norteamericanos y europeos: Aportes para una convivencia armónica local* (Informe final). Cuenca: Avance Consultora.

Andrews, M. (2019, July 18). Dream of retiring abroad? The reality: Medicare doesn't travel well. *New York Times.* https://www.nytimes.com/2019/07/18/business/medicare-retire-abroad.html. Accessed 19 July 2019.

Bickel, D. (2017). 'Missing' friend found dead in his home. *Mexico News Daily.* https://mexiconewsdaily.com/opinion/missing-friend-found-dead-in-his-home/. Accessed 26th July 2019.

Bustamante, A. V., Laugesen, M., Caban, M., & Rosenau, P. (2012). United States-Mexico cross-border health insurance initiatives: *Salud Migrante and Medicare in Mexico. Revista Panamerica de Salud Pública. SciFLO Public Health, 31*(1), 74–80. https://www.scielosp.org/article/rpsp/2012.v31n1/74-80/en/. Accessed July 26th, 2019.

Castro, R. (2014). Health care delivery system: Mexico. In *The Wiley Blackwell Encyclopedia of health illness, behavior and society* (pp. 836–842). Chichester: Wiley.

Gokhale, K. (2013). Heart surgery in India for $1,583 Costs $106,385 in U.S. *Bloomberg Business News*. https://www.bloomberg.com/news/articles/2013-07-28/heart-surgery-in-india-for-1-583-costs-106-385-in-u-s-. Accessed 19 July 2019.

Government of Mexico. (2019). *Reform of Seguro Popular will contribute to the creation of a national health system*. https://www.gob.mx/salud/en/articulos/reform-of-seguro-popular-will-contribute-to-the-creation-of-a-national-health-system?idiom=en. Accessed 5 July 2019.

La Parra, D., & Mateo, M. A. (2008). Health status and access to health care of British nationals living on the Costa Blanca, Spain. *Ageing & Society, 28*, 85–102.

Lucio, R., Villacres, N., & Henriquez, R. (2011). Health system of Ecuador. *Salud Pública de México, 54*(2S), 177–187.

Miller, L. J., & Lu, W. (2018). These are the economies with the most (and Least) efficient health care. *Bloomberg News, Business*. https://www.bloomberg.com/news/articles/2018-09-19/u-s-near-bottom-of-health-index-hong-kong-and-singapore-at-top. Accessed 8 Jan 2019.

Morales, A., Miranda,P., & Zavala, M. (2019). *AMLO anuncia sustituto al Seguro Popular. El Universal*. https://www.eluniversal.com.mx/nacion/amlo-anuncia-sustituto-al-seguro-popular. Accessed 5 July 2019.

National Assembly of Ecuador. (2008). *Constitution of the Republic of Ecuador*. Article 37, Section 1, Chapter 3; Article 358–365, Section 2, Chapter 1. Translated version. http://pdba.georgetown.edu/Constitutions/Ecuador/english08.html. Accessed 8th Jan 2019.

OECD. (2016). *OECD reviews of health systems: Mexico 2016*. Paris: OECD Publishing.

Pan American Health Organization. (2017). Health in the Americas+, 2017 edition. In *Summary: Regional outlook and country profiles*. Washington, DC: PAHO.

Sloane, P. D., Cohen, L. W., Haac, B. E., & Zimmerman, S. (2013). Health care experiences of U.S. retirees living in Mexico and Panama: A qualitative study. *BMC Health Services Research, 13*, 411.

Villacres, T., & Mena, A. C. (2017). Payment and financial resources management for the consolidation of Ecuador's health system: Mechanisms of payment and management of financial resources for the consolidation of the health system of Ecuador. *Pan American Journal of Public Health, 41*(3).

# Chapter 7
# Long-Term Care Options for Retired Americans in Cuenca, Ecuador and San Miguel de Allende, Mexico

Philip D. Sloane and Sheryl Zimmerman

The overwhelming majority of persons who retire internationally do so when they are healthy. What happens when they develop enough disability to require long-term care services is largely unknown (Warnes 2009). The few existing reports are of European international retirees, as crossing borders in retirement began in large numbers earlier in that continent than in the Americas. Reports from Spain identified a relative lack of post-hospital care and of nursing home resources, such that retirees needed to turn to non-family networks of fellow migrants, to charitable organizations, and to paying out of pocket more than their counterparts back in Britain (Haas 2013; Hall and Hardill 2016). Furthermore, retiree interviews in Spain indicated that much planning is needed but is not always done, and that, as a result, some members of the international retirement community fall through the gaps and do not obtain needed services (Hall and Hardill 2016). In contrast, the development of private residential care facilities specifically catering to Germans has been described in Greece, Eastern Europe, and Thailand, with payment provided by the German government, but accompanied by concerns that quality in some of these low-cost settings may be substandard (Ormond and Toyota 2016).

In the United States, 35% of persons aged 65 and older report one or more disabling conditions, 9% have cognitive impairment, and 8% need help performing daily activities (Houser et al. 2018). The majority of long-term care services in the U.S. are provided informally by nearly 40 million family caregivers; thus, nearly 13% of the general population, and a far higher proportion of older persons, are providing long-term care services (AARP Public Policy Institute 2015). However, many individuals with disabling conditions either do not have family or require more assistance than family can provide, and for them a variety of formal (paid) services are available. In the U.S., formal long-term care services are provided by

P. D. Sloane (✉) · S. Zimmerman
University of North Carolina at Chapel Hill, Chapel Hill, NC, USA
e-mail: philip_sloane@med.unc.edu

© Springer Nature Switzerland AG 2020                                          123
P. D. Sloane et al. (eds.), *Retirement Migration from the U.S. to Latin American Colonial Cities*, International Perspectives on Aging 27,
https://doi.org/10.1007/978-3-030-33543-4_7

an estimated 1,888,483 home health or personal care aides, 15,572 nursing homes, 28,900 assisted living communities, and 4600 adult day services centers. Numerically, these figures break down to 252 home health or personal care aides, 28 nursing home residents, 17 assisted living residents, and 0.6 adult day services users per 1000 persons aged 65 and older (Houser et al. 2018; US Centers for Disease Control and Prevention, National Center for Health Statistics 2016; US Centers for Disease Control and Prevention, National Center for Health Statistics 2016).

Given these figures, it is reasonable to assume that a sizeable amount of long-term care service use exists among the approximately 8000 and 17,000 retirees from the U.S. and Canada who respectively are living in Cuenca, Ecuador and San Miguel de Allende, Mexico. However, deriving an estimate of need is challenging, because overall statistics are lacking. We assume that need is less than the proportions noted above, because on average the retirees in these cities are younger than those in more established retirement settings in the U.S. Additionally, it is well known that persons who relocate in retirement tend to be healthier and better off economically than their peers who stay put, both of which are factors associated with low nursing home use. For example, Florida – the most popular relocation retirement state within the U.S. – has only 24 nursing home beds per 1000 persons aged 65 and older, whereas in Iowa the rate is 66 per 1000. Still, we anticipated that some of the estimated 25,000 retirees across these two cities had to be using long-term care services. Indeed, development of formal long-term care services is crucial for a community's success as a retirement destination.

The cost of long-term care services in the U.S. is exorbitant, so, we surmised that the potential to obtain equivalent services at a lower cost could attract retirees to Latin American cities. Indeed, as early as 2007 a front-page article in *USA Today* had written about a "small but steadily growing number of Americans who are moving across the border to nursing homes in Mexico, where the sun is bright, and the living is cheap." The article went on to describe a 74 year-old American in Ajijic, Mexico, who was receiving assisted living for a quarter of what she would have paid back home. That setting, according to the report, included "a studio apartment, three meals a day, laundry and cleaning service, and 24-h care from an attentive staff, many of whom speak English."(Hawley 2007)

We also knew, however, that formal long-term care services are rare in Latin America. Far fewer than 1% of the region's population over age 60 lives in residential settings, and nursing homes and assisted living settings are practically nonexistent. Formal home care providers are largely rare as well, and what formal care does exist is concentrated in families with high incomes. Hence, for the vast majority, care is provided by family members. Quality is uneven and often leads to stress on family budgets and relationships (Caruso et al. 2017).

This lack of formal services is accompanied by a relative absence of regulation, and, when long-term care services are provided, it is by the public sector – a common phenomenon in lower and middle income countries (Caruso et al. 2017). Yet much of the developing world has rapidly-growing senior populations, and in response, the World Health Organization recently identified long-term care as a significant policy priority for the developing world (Pot et al. 2018). However, at the governmental level it is hard to prioritize long-term care, when primary care, emergency services, and public health still need development (Norori 2018).

Indeed, much of Latin America resembles the U.S. in the early 1900's, where most residential long-term care settings were "poor houses" or "county homes" for indigent persons who had no family to care for them. In Mexico, each community tends to have one of these *asilos*, which serve older persons who have no family and are supported by donations. In that context, we wondered whether retirees from the U.S., Canada, and Europe might constitute a force that would foster development of formal long-term care services, most likely in the private sector.

In 2008 we had visited several residential long-term care settings in Ajijic and San Miguel de Allende, Mexico. One was privately owned and included several adjacent houses, each of which provided room and board, 24-hour staff, nursing care as needed, and visits by private physicians. The other two were *asilos* that were developing apartments to attract retirees from North American and Europe as paying customers, as a means of augmenting resources for the residents (Sloane and Zimmerman 2007). Thus, while what we had observed was not nearly as widespread as might have anticipated from media reports, it did appear that retirees from the U.S. and other developed countries were indeed stimulating some long-term care service development. Because of these observations, in 2014 we predicted that international long-term care would grow rapidly in the coming decades (Sloane et al. 2014). Furthermore, we anticipated that much of the long-term care services we'd encounter would be private care in the home, as many retirees from the U.S. in Mexico and Panama who we had interviewed in a previous study said that if disabled they would stay at home with personal care services, because such services were widely available and quite affordable (Sloane et al. 2013).

Therefore, in our study team's site visits to Cuenca, Ecuador, and San Miguel de Allende, Mexico, we wanted to take the pulse of the development of both home and residential long-term care services in our study cities. We did not have the time or resources to develop a census; instead we used a combination of survey results and on-site interviews to describe retiree attitudes and available resources. As part of our online survey of expat retirees (see Appendix for details), we asked respondents about the location and type of services they planned to use if and when they became disabled. Additionally, we used a snowball informant technique to identify and then visit and interview in depth at least three retired Americans or Europeans in each city who were using formal long-term care services on a daily basis. This chapter describes the results of these investigations.

## 7.1 Attitudes of Survey Respondents

Our online survey included 667 persons who responded to the item "I plan to live in [Cuenca or San Miguel de Allende] even if I became disabled." Of the respondents, the majority – 56% in Cuenca and 52% in San Miguel de Allende – either strongly agreed or agreed that they would remain. Another large group – 36% in Cuenca and

33% in San Miguel de Allende – reflected uncertainty, responding "somewhat agree", "neither agree nor disagree", or "somewhat disagree" to the item. Only a small minority – 10% in both Cuenca and San Miguel de Allende – disagreed or strongly disagreed with the statement. These results are displayed in Fig. 7.1.

To better understand some of the factors influencing retiree responses regarding where they would prefer to receive care if disabled, we analyzed the relationship between retiree characteristics and responses to the item "I plan to live in [Cuenca or San Miguel de Allende] even if I became disabled." Characteristics we considered were age, education, income, whether or not they were living with a spouse/partner, year of residence in the city, amount of time spent in the city in the previous year, the number of times they visited their home country in the previous year, where they kept the majority of their savings, the city (Cuenca or San Miguel de Allende), and their primary reason for immigrating internationally.

Table 7.1 displays the results of the multivariable regression analysis. They indicate that persons who plan to live in Cuenca or San Miguel de Allende even if disabled are significantly more likely to be older, to already get most or all of their health care in the city, and to have migrated because of a desire to get away from their home country. It also shows a strong though not statistically significant trend for expats to plan on returning to their home country if they have a higher monthly income and/or returned to their country of origin two or more times during the previous year. Education, years of residence, location of their savings, and whether single or in a coupled relationship were factors that, while associated in bivariate analyses (not displayed) were not associated in the multivariable analysis with desire to remain in the country if disabled.

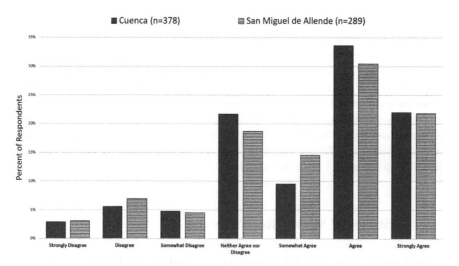

**Fig. 7.1** Responses of Immigrant Retirees in Cuenca and San Miguel De Allende (N = 662) to Survey Item "I plan to continue to live in [city] even if I became disabled"

**Table 7.1** Relationship between retiree characteristics and plans to continue living in city even if disabled[a]

| Variable | Coefficient | Standard error | P-Value |
|---|---|---|---|
| Older age | .038 | .018 | 0.033[b] |
| More years of education | −.002 | .267 | 0.995 |
| Higher monthly income | −.415 | .225 | 0.065 |
| Living situation | −.238 | .231 | 0.303 |
| Years of residence in [CITY] | .035 | .024 | 0.142 |
| Months in past year spent in [CITY] | | | |
|    9 or fewer months | Reference | | |
|    10–11 months | .163 | .274 | 0.551 |
|    12 months | .320 | .274 | 0.242 |
| Number times visited home country past year | | | |
|    None | Reference | | |
|    Once | −.113 | .268 | 0.673 |
|    Two or more visits | −.525 | .284 | 0.065 |
| Location of majority of savings | .408 | .261 | 0.119 |
| Received higher proportion of health care in destination country (Mexico or Ecuador) in previous year | .006 | .003 | 0.035[b] |
| Primary reason for immigration in retirement | | | |
|    Affordable retirement or health care | Reference | | |
|    Amenities (lifestyle, climate, culture, adventure) | .150 | .238 | 0.527 |
|    Get away from home country | 1.422 | .482 | 0.003[b] |
|    Other | .213 | .440 | 0.628 |

[a]Ordered logistic regression with outcome variable trichotomized as "disagree," "undecided" and "agree." Number of respondents analyzed in model = 489, 285 who were living in Cuenca and 204 who were living in San Miguel de Allende
[b]Statistically significant at p ≤ 0.05

While acknowledging that the vast majority of expats surveyed did not plan to return to their native country if disabled, our research staff, two of whom were social workers, expressed skepticism. One wrote this in her field notes:

*"[a couple we interviewed] plan to live out the rest of their lives in San Miguel de Allende. However, I felt they were unrealistic about what possibilities may bring them back to the US. Their daughter recently married, and grandchildren are in the near future, but they have no plans to follow their children and grandchildren around. However, they took care of [the wife's] father for years; he had advanced Parkinson's disease and lived far into his 90's. If one of them has a chronic disease like this, and the other is impaired or has died, I would think one of their two kids would want to move them back to the states, especially after flying to Mexico every month for a while because of hospitalizations, chemotherapy, multiple falls, and many other possible issues that could arise in chronic illness."*

## 7.2   Challenges Locating Persons with Disabilities

Considering that the majority of retirees expressed a desire to remain if disabled, we were surprised at how difficult it was to locate persons with disabilities. The retirees with whom we first came into contact, who were active on the internet and/or frequented popular expat meeting locations, tended to draw a blank when we asked about persons with significant disabilities. This comment from our field notes is illustrative:

> *"A remarkable thing is that, in spite of being in Cuenca for 8 years and widely networked, [a married couple] could not identify a single person who was receiving disability services. There were some around, they noted, but these were not in their circle of friends. 'I'm about to turn 70,' [the man] said. 'I feel good.' Their 4th floor, walkup only, penthouse apartment was a testimony to their current vigor."*

One reason for this lack of familiarity may be that few expats with disabling conditions live in these cities. However, another possible reason is isolation due to poor access to public areas. Both countries, and indeed much of Latin America, lack the regulations of more developed countries requiring disability access to all public spaces. Similarly, street-level entry or wheelchair ramps are quite rare among businesses in both cities, and, as reflected in the field note above, many apartments have stair access, requiring significant mobility.

In Cuenca we spoke with a woman in a wheelchair at the Sunrise café, a common expat meeting place. She said that accessibility had been improving in recent years, with more ramps and handicapped bathrooms cropping up. She attributed the growth of disability awareness in Ecuador to the fact that their current president, Lenin Moreno, used a wheelchair, having been disabled years ago when he was shot during a holdup in a grocery store parking lot (McBride 2017). She noted, however, that some ramps are poorly designed and don't actually serve their intended purpose. She had, for example, encountered wheelchair ramps that didn't come all the way up to floor level or bathrooms that didn't have enough room to accommodate a wheelchair.

San Miguel de Allende is especially challenging for persons with disabilities. The majority of streets are of uneven cobblestones. Sidewalks are narrow and of cobblestones as well, though their surfaces are more even than the streets. Curbs tend to be high and wheelchair access ramps nonexistent in the streets and rare at building entries. An exception is the *bibilioteca* (library), a favorite meeting place for the expat community, which is largely supported by expat donations. Next to its two entry steps, the building has a ramp and railing, thereby providing ground-floor access for wheelchairs and walkers, although access to the second floor, which houses the café, is by stairs only.

With much networking among expats, we did eventually locate several persons in each city who were regularly accessing both home-based and residential long-term care services. We also located and met with several home care providers in each city and identified and site-visited residential long-term care settings in both sites.

## 7.3   Home Care Services for Persons with Disabilities

Home care is the preferred long-term care option of most expats in both Cuenca and San Miguel de Allende – a finding that echoes what we had found in our earlier study of retirees in Mexico and Panama (Sloane et al. 2013). One retiree, who had arranged 24-hour home care for his wife, put it this way: *"Home health care is like the Cadillac of care, where you have someone who is totally dedicated to the care of a particular person."* For most expats, however, the desire to have home care is theoretical rather than based on knowledge of the home care options or of the organizational skills required to obtain and supervise home services. *"The long-term care thing is a potential concern,"* was all one interviewee could say in response to questions about what he would do if no longer independent. On prompting, he added, almost as an afterthought, *"we would hire a live-in individual."*

Such an individual would ideally be what is sometimes referred to in the U.S. as a "universal worker' – someone who can provide:

- nursing services such as medication administration, vital signs monitoring, skin care, and supervision and assistance with preventive physical and cognitive exercises;
- assistance with activities of daily living, such as help with bathing, dressing, grooming, eating, and using the toilet; and
- housekeeping services such as cleaning and cooking.

The amount and type of services would depend on the needs of the older person or older couple.

Finding such a worker is not easy. For one thing, role fragmentation is common in many countries, as in the U.S.. One expat in Ecuador put it this way:

> *"The nurse will say to you, you can't make me cook, I'm a nurse. And if you're a cook, you can't ask them to do nursing….we don't have that combination."*

Despite this tradition of job segmentation, we found several expats who had arranged for individuals with nursing training to work full-time at a job that included most or all of the above universal worker tasks. Both Cuenca and San Miguel de Allende clearly have some individuals interested in this type of work, although we were unable to fully explore what kinds of training these "nurses" had or the size of the available pool of nurses for home care. We do know that having nurses highly involved in home-based care was common in the U.S. a century ago, when hospitals tended to rely on student nurses, and institutional long-term care was not yet developed, leading to a surplus of trained nurses compared to available hospital positions. Indeed, in the initial decades of the twentieth century, most U.S. nurses did private duty work, often having to use their own resources to secure employment, until over time private duty agencies arose to connect nurses with prospective clients (Whelan 2012).

## 7.3.1 Home Care in Cuenca

In central North Carolina, where the authors live, there are dozens of licensed home care agencies. In contrast, finding and coordinating home care services is more difficult in Cuenca, due to a relative lack of formal home care agencies. Consequently, informal networks of expats helping expats were providing much home care, including a tendency of single persons to share apartments as they began to need help. When informal support was not sufficient, finding a home care provider involved considerable networking, often using a facilitator. As one expat stated, *"There's really no one doing [home care placement] on a commercial basis here."*

For native Cuencanos, arranging home care is easier, according to staff member at the Ministry of Economic and Social Inclusion (Ministerio de Inclusión Económica y Social). He said that Cuenca has geriatric centers that pair older Ecuadorians with caretakers, but that the service doesn't work with expat retirees because the majority of caretakers don't speak English.

With networking, however, home care could be found, and all of the individuals and couples we met who were using home care services praised their caregivers for both their high quality and their low cost. This quote from an expat who paid for full-time home care for his wife who has dementia is typical:

> *"We've had wonderful help here actually. The dementia care (includes) bathing, cooking, cleaning, memory care, exercise, physical therapy, mental puzzles, working to keep the brain working."*

Furthermore, the cost of home care is much less than it would be in the U.S., where government support for extensive home care is lacking. One couple estimated 24-hour home care in Florida at $14,000 a month. In Ecuador, they were paying $3500. *"It's a lot of money for someone to pay out of pocket,"* the husband said, *"but it sure isn't $14,000."*

Another expat had brought her 90+ year-old mother to join her in Cuenca, because she felt that the home care there would be superior to anything her mother could afford in the States. Organizing care for her mother took a lot of effort and coordination, however. A key component was a live-in nurse she had hired full time, complemented by several other workers to provide her with time off:

> *"I pay her room and board and 600 dollars a month. And people told me I was paying too much, if you can imagine. She's a fully trained RN and has a physical therapy degree. And because Mom got more sick, I paid her $800 a month. I increased her pay because I just felt like she, she was doing so much more than what I wanted, than what I had told her was going to be necessary. And so anyway, she gets time off when A-- the housecleaner you met – comes; and she gets time off three days a week when B—comes, and then she gets Saturday and Sunday nights off when another woman C--- comes. Or sometimes I do those nights. And often I'm here with Mom when the nurses are here and we're doing things together."*

The existence of the large immigrant retiree community represents a business opportunity, according to the owners of VIP Health, a new company that exclusively

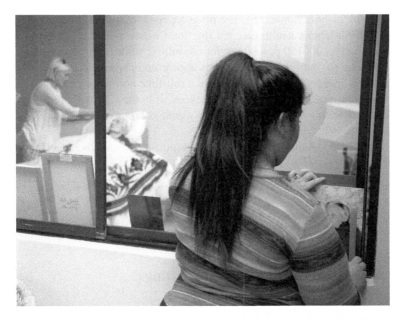

**Fig. 7.2** This retired American brought her 92 year-old mother to live with her Cuenca. The 24-hour home care she arranged includes a full-time registered nurse (foreground)

serves English- speaking retirees. Their services include regular home health care and medical appointment arranging, and at the time of our interview they were working on implementing emergency services. The co-owners, a nurse and a businessman, had immigrated several years earlier from India. After their initial job opportunities did not work out, they identified and sought to address a need for English-speaking post-hospital home care. Apparently, Ecuadorian hospitals do little or no discharge planning, depending instead on families to provide or obtain the needed services; therefore, the nurse began providing post-hospitalization home care to expats and after a year joined with her partner to form the company. The new company viewed the biggest opportunities to be post-acute care after hip surgery, knee surgery, and cosmetic surgery. The owners noted that few expats want to go to the few assisted living settings in Cuenca, and that local doctors make home visits, which makes home care more feasible (Fig. 7.2).

## 7.3.2   Home Care in San Miguel de Allende

San Miguel de Allende, with its larger, more established expat retiree community, has better-developed formal home care resources for immigrant retirees. Several agencies are available to arrange home care, although the cost of care is higher than hiring privately. One company whose owners we interviewed charges $4000–$6000 per month for 24-hour home care. The price depends mostly on the type of professional involved, from a health care aid (the least expensive) to a registered nurse. They also charge 100–150 pesos/hour for consulting work.

For persons paying privately, the costs appear comparable to those in Cuenca. One relatively independent retiree who had a full-time non-nurse as a care provider and companion, was paying him $300 per week. If in the U.S. or Europe, she said, she would *"probably have somebody who would come in once week to check on me,"* due to the price difference.

Another expat, who employed a full-time non-nurse caregiver for his spouse with dementia, said:

> *"Officially the minimum wage is about 60-70 pesos a day. You won't pay that to a caregiver, but you might pay the equivalent of $25/day per person to cover 24/7. That's feasible."*

## 7.4 Residential Long-Term Care Settings

Throughout Latin America, residential settings providing long-term care are present but rare. Traditionally, homes for older persons have been charities for destitute individuals without families, operated with support from the local government and/or the church. Standards of care vary, and resources are always tight. Recently, however, some changes have begun to occur, particularly in the private sector.

As part of our research in Cuenca and San Miguel de Allende, we searched for residential long-term care settings that were housing one or more expat retirees. The ones we identified fall into three categories: traditional homes for indigent older persons (*hogares para ancianos* or *asilos*) with apartments for immigrant retirees; private facilities primarily for local residents; and private facilities catering primarily to immigrant retirees.

### 7.4.1 Homes for Indigent Older Persons, with a Private Pay Option

As noted previously, homes for older persons, similar to the old U.S. county homes and poor houses, are operated in many Latin American communities. Such a home is operating in San Miguel de Allende and has begun to accept expat retirees as a way of augmenting its income. It is called *Alma*.

*Alma* was founded by an American but is now managed entirely by Mexicans. We first visited it in 2006. At that time it housed no Americans, only Mexicans, but the administrator was thinking that appealing to immigrant retirees could improve Alma's budget, since the organization was entirely dependent on donations and fundraising efforts (Sloane and Zimmerman 2007). In 2018, when we re-visited Alma, it housed seven expat retirees, each of whom was paying $14,000 pesos ($760 U.S. dollars) a month for a private room. To serve its 41 residents, Alma employed a staff of 19. Of these, two were administrative staff, and the remainder provided personal care, housekeeping, gardening and meals. A doctor visited 3 days a week; a nurse was on duty 6 days a week; and a physical therapist came 3 days a week. In the U.S., similar residential care would cost thousands of dollars.

Most rooms are dormitory style, with four persons per room. Activities are few, provided largely by volunteers, and there is a pet dog and cat on site. Interestingly, Alma had recently changed its policy to only let foreign retirees become residents if they had a family member who could take responsibility in the event their health deteriorates, or they could no longer pay for the care. This requirement reflects the fact that no option currently exists in Mexico (or Ecuador) to assume responsibility for persons who require a change of status because of health or financial reasons and cannot make decisions for themselves.

## 7.4.2 Private Assisted Living-Type Facilities Primarily for Local Residents

In Cuenca, the only residential long-term care options we identified that expat retirees were using consisted of two private assisted living-type facilities that primarily served the local population. We only found four retirees using these settings. Both facilities were operated by Ecuadorians, mainly served old and disabled local residents, and had staff who almost exclusively spoke Spanish – the exception being the physicians, who in both cases did speak some English. The two facilities we identified were called *Los Jardines* and *Años Dorados*.

*Los Jardines* is a modest private assisted living community near the center of Cuenca. Largely serving old and disabled Cuencanos, its residents also included a few immigrant retirees. Due to one of the foreign resident's multiple surgeries and need for long term health care, she reported that she had been declined further use of Ecuador's public healthcare insurance. She was currently paying out of pocket for the facility while trying to identify and purchase a private health care plan. (For a discussion of health insurance options in Ecuador and Mexico, see Chap. 6).

The other setting we visited, *Años Dorados*, is located in a semi-rural area about 15 miles downstream along a valley that has recently seen much building, due in part to it being accessible in 20 minutes to central Cuenca by a limited-access highway. *Años Dorados* has two sites. The larger was a sprawling, 2-story, U-shaped building with 14 occupants – one American, the remainder Ecuadorian. The smaller was a 2-story house with 4 occupants, one of whom was an American. The director, a gerontologist, was quite in tune with current thinking about care of seniors, such as person-centered care, socialization, activity therapy, and minimizing medications. The physician visited 3 times a week and was available to make on-site visits

**Fig. 7.3** Outdoor area at Años Dorados in Ecuador. The curved object in the center is an oven used for making traditional Ecuadorian dishes; the building in view is the administrative area; resident areas are in building to the right

24 hours a day, and a nurse was on site daily. As a result, *Años Dorados* could manage many acute conditions on site, such as pneumonia, and was able to provide intravenous fluids, parenteral medication, and respiratory therapy. The facility, and others of its type, must conform to requirements of both the Ministerio de Inclusión Económica y Social (MIES) and the Ministerio de Salud. Charges for a single room with private bath are $800–$900 per month. If a resident needs an intensive level of service, a $50/month additional fee is charged (Fig. 7.3).

In San Miguel de Allende, we identified an assisted living-type facility that housed a number of retired expats. Its name is *Santa Sofia Casa de Reposo*, and it is located in Guadiana – one of the neighborhoods most heavily populated with retirees from the U.S. and Canada. The price for single room is 40,000 pesos/ month (2000 U.S. dollars); for a shared room it is 35,000 pesos/ month). The price includes room and board; around-the-clock staff availability; activities such as tai chi, music, and art; chronic disease care; post-operative care; and acute care for emergencies. When we visited, two of the facility's 17 residents were retired Americans (Fig. 7.4).

**Fig. 7.4** Semi-private room in Santa Sofia Casa de Reposo, an assisted living community in San Miguel de Allende catering largely to retired expats. Monthly cost for room and board, personal care, and 24-hour staff availability in a semiprivate room in 2018 was $1892 US dollars

### 7.4.3 Private Assisted Living Facilities Catering Primarily to Expat Retirees from the US or Canada

In Cuenca there are no private long-term care facilities that primarily target expat retirees. We did hear about one on the coast that had opened a number of years beforehand with the intention of attracting expat retirees. It was small – with room for only eight or ten persons. Even at that size, the facility could not attract enough clients, and after some time it closed. *"They could never fill them because the Gringos wouldn't pay the price, even though it's far cheaper than the States,"* our informant commented. We also heard of another facility in Ecuador that was catering to expats. It's in Vilcabamba – a valley with a reputation as a cradle of longevity, that has been nearly overrun by immigrants from the U.S., Canada and Europe, many of whom are retirees. A municipal-sponsored home for older adults, reportedly charging $1000 per month, could be a prototype of the kind of resource that might appeal to retirees. However, as of December 2018, their website did not advertise in English and, therefore, did not appear to be primarily targeting retired Americans.

In San Miguel de Allende, however, with its larger, more established, retiree population, we did identify a long-term care setting that catered almost exclusively to expat retirees. It is *Cielito Lindo*. Located about 6 miles outside of town, it is embedded in an upscale, gated community called *Los Labradores*, which ironically translates as "the farmers." *Los Labradores* was developed and is managed by a company from Toronto, Canada. It is in many ways like a continuing care retirement community in the states, with approximately 100 homes, a clubhouse, and activities for independent older persons, plus Cielito Lindo – an assisted living community that provides care until death.

When we visited *Cielito Lindo* 18 of its 20 beds were occupied. Of the residents, eight were described as "high need" and 10 as "more independent," although all the residents we saw appeared quite impaired. All but two were expats, and all paid the same – almost $2900 U.S. dollars per month. Interestingly, three had run out of money, a situation that creates a quandary for long-term care providers in Latin America, since there is nowhere to send such an individual, and putting them out on the street would be a public relations disaster. The facility had a full time doctor – an internist with geriatric training, who spent some time in Cleveland Clinic. He is in the facility from 9 am until 3 PM on weekdays, and in the afternoons does consultations in the *Los Labradores* community (by appointment). Hospitalization is at a private hospital in San Miguel de Allende.

## 7.5 The Disability Dilemma for Retired Expats: Stay or Return?

Relocation in retirement can be divided into three categories: amenity moves, disability moves, and residential long-term care moves (Table 7.2) (Litwak and Longino Jr. 1987).

- Amenity moves are typically done by younger, healthy retirees, who are seeking quality of life; probably 99% of the expats who moved in retirement to Cuenca and San Miguel de Allende were making amenity moves.
- Disability moves typically involve relocating closer to family and health care resources. For example, the woman in photograph 7.1 made such a move, having been brought to Cuenca by her daughter. Our survey of retirees in both sites suggests that the majority do not anticipate moving back to their home country if disabled; however, we can't help wondering if more will and do move than anticipate doing so. For example, it is quite possible that after two or three trips abroad to check up on a mother or father's hospitalization, a son or daughter may begin exerting pressure to have their parent move closer.
- Moving to a residential long-term care setting is clearly something that retirees in Mexico and Ecuador don't want to make. They are reassured by the availability of modest 24-hour home care, which is highly praised by those who receive it (Sloane et al. 2013). Furthermore, Latin America has a large reservoir of underemployed middle-aged women, some with nursing training, who could be employed in an expanded home care industry.

On the other hand, decision-making in advanced age involves different factors than early in retirement. For one thing executive function – the ability to do things

**Table 7.2** The three types of moves that retirees typically make

| Type of move | Who makes the move | Reason for move |
| --- | --- | --- |
| Amenity move | Younger retirees who are relatively healthy, have at least some financial resources, and (less so now than in the past) have a partner who moves with them | Improve quality of life; have access to more leisure activities and/or a more attractive environment |
| Disability move | People who have moderate disability | Be closer to social support from family and/or health care services |
| Residential long-term care move | Persons with significant disability who lack adequate cognitive and/or family resources to manage at home | Relocate to assisted living or nursing home |

Adapted from Litwak and LonginoLitwak and Longino Jr. (1987)

such as plan and coordinate care – is frequently impaired in persons with significant chronic illness, especially if they are over 80. Indeed, impaired executive function appears to be a key factor in entry to U.S. assisted living settings (Kaufer et al. 2008). So, inability to coordinate one's own home care and manage a household may well lead many who prefer home care to relocate to an assisted living or nursing home setting, especially if they don't have a spouse or family member to provide the needed coordination. The emergence of assisted living-type settings catering to expats in San Miguel de Allende indicates that indeed this is sometimes the case (Fig. 7.5).

Payment and insurance are an important ingredient to these issues. In the US and Canada, and increasingly in some European countries as well, long-term care services are expensive and not well covered by insurance, but if someone spends down their resources, they can have long-term care paid for by a safety-net insurance program. In the U.S. this insurance program is Medicaid, and nationally about half of nursing home residents are on Medicaid. No safety net of this kind exists in Latin America; so, the best an expat can hope for is to either to not outlive their resources, to find an asilo that will take them in, or to return to the U.S. and qualify for Medicaid. Any such move requires coordination, and for those with family, that coordination is quite likely to bring them back to their home country.

What does this mean for the retirees living abroad and the hundreds of thousands more who may move abroad in the coming decade? The vast majority expect to be in the amenity stage for many years and, even as they age, often find it difficult to plan concretely for dependency. Others, perhaps more sanguine, acknowledge that another move may be in their future. As one interviewee put it:

> "I'm 68, in excellent health. I can expect to live into my 90s. I have Medicare... Say I got breast cancer and I needed chemo and radiation and I needed it covered by Medicare....Any conditions that would be kind of expensive like that, that I couldn't afford here really. But I might look into it and see if I could have the same treatments here and afford to have it here, (in which case) I might not go back. It's one of those things where, cross that bridge when you come to it."

**Fig. 7.5** Staff member at Santa Sofia Casa de Reposo assisting a resident safely obtain exercise and exposure to sunlight. Interviewees in both cities consistently praised the care provided in home and residential settings

## 7.6 Current and Future Role of Long-Term Care Services in Cuenca and San Miguel de Allende

Our research located few retirees in either Cuenca or San Miguel de Allende who were using formal long-term care services. Several factors could contribute to this phenomenon:

- The retiree community needs more time to age into a need for long-term care. Retirees typically relocate when relatively young, with ages 55–65 being the peak relocation ages. This situation means that the average retiree is likely to have 15–20 years of active life expectancy before needing long-term care services.
- Many retirees employ a housekeeper who helps with errands, cleaning, cooking and other chores. It is possible that over time some such individuals begin to carry out long-term care functions, such as assistance with dressing and grooming, without the arrangement being thought of as long-term care.
- Retirees who are receiving long-term care services at home may be confined to home and therefore relatively invisible to the local community and their fellow expats.

- When confronted with the reality of needing long-term care, more retirees may return to their home country than had anticipated doing so. After all, entry to long-term care is rarely smooth; instead it is often preceded or accompanied by an accelerating need for health services, often with several hospitalizations, and family back home could pressure the individual to move closer. For example, it would not take many urgent long-distance trips to visit a sick parent to convert a daughter, son, or grandchild into a strident advocate for relocation "home."

Currently, formal long-term care services are underdeveloped in both Cuenca and San Miguel de Allende. However, it appears that the foreign retirees are helping catalyze the development of such services. Thus, we found nascent home care agencies in both cities, where the tradition has been family care augmented, if necessary, by informally contracting with someone to provide home-based services. Similarly, we found that assisted living-type care programs were just beginning to develop, and that housing of immigrant retirees with the ability to pay was often part of the facility's plan.

These developments, if they continue, will constitute an important contribution of the expat retirees to their destination communities, as it is widely acknowledged in policy circles that Latin America has a pressing need to develop long-term care services due to rapid population aging and high prevalence of chronic, disabling conditions in the region. Development of such services can be a positive force for the local economy; in fact, one report concluded that investing in health services would be better for the economy than investing in construction (De Henau et al. 2017). Latin America's approach to long-term care may, however, place more emphasis on subsidized home-based caregiving, much of which is delivered by members of the extended family, and less emphasis on residential settings such as nursing homes (Pot et al. 2018).

The World Health Organization's global strategy and action plan on ageing and health noted that middle and low-income countries need to "build on the strengths and cultural values of existing family care and enable a partnership between families and other caregivers/providers to ensure sustainability and equity. Long-term care systems are not intended to replace existing care frameworks or approaches, such as rehabilitation or palliative care, but rather serve as a means of unifying and augmenting existing care approaches for care-dependent people." Even developing countries are adopting public-private cost-sharing models of long-term care, something that will also be needed in Latin America (Pot et al. 2018).

Indeed, most Latin American countries need policies regarding the development of long-term care services, the form that they will take, and the role of national government in fostering long-term care. However, in many countries, poverty and general medical care are such pressing issues that long-term care is not likely to emerge as a policy priority for some time (Norori 2018).

One issue in the development of long-term care services is whether or not medical services should be delivered separately from other supportive services. In home care, for example, is it really appropriate for a person with a disabling medical condition to require a nurse for medical services, an aide for help with activities of daily living, and a housekeeper for such services as cleaning, laundry, and cooking? (Caruso et al. 2017). Several individuals in the expat community spoke with us that this traditional subdivision of tasks by profession was a barrier to cost-effective home-based care. On the other hand, organized professionals are often very effective lobbying groups; so how long-term care will evolve is uncertain.

Another issue is the growing prevalence of Alzheimer's disease and other dementias, which are expected to accompany the aging of the population of Latin America (Custodia et al. 2017). These neurodegenerative diseases create an ongoing, increasing care burden over many years, while at the same time rendering the individual incapable of managing their own affairs, as well as being accompanied by challenging behaviors. For these reasons, dementia is the most common diagnosis leading to failure of home care resources and consequent placement in a nursing home in developed countries with well-developed home care resources, such as the United States and the Netherlands. Thus, it is quite likely that more formal services will need to be developed in Latin America to relieve families of the relentless demands of dementia care.

Insurance complicates these issues. In the U.S., Medicare does not pay for long-term care services, but Medicaid does. Long-term care insurance is being aggressively promoted, but many policies have exclusions, including home care and any type of international care. As one of our interviewees said:

> "[In the United States], if I have 24/7 home care, it would be even more than $10,000 dollars a month. But here, I can have homecare for, like, four thousand a month. And I have long term care insurance that covers international care, but it has a limit. So, if I was living in the United States... I would go through my long-term care insurance in about two years, two or three years. But my disease can last 15 years. So, I'm coming where I can afford to have my care provided by the insurance longer."

In conclusion, it is hard to predict whether and to what extent formal long-term care services in home and institutional settings will develop in Latin America. However, the early service use patterns observed in this study suggest that, if retirement migration continues to grow, we can anticipate considerable expansion of long-term care services, with a continued emphasis on home care.

# References

AARP Public Policy Institute. (2015). *2015 report: Caregiving in the U.S.* https://www.aarp.org/content/dam/aarp/ppi/2015/caregiving-in-the-united-states-2015-report-revised.pdf. Accessed 22 June 2019.

Caruso, M., Galiana, S., & Ibarrarán, P. (2017). *Long-term care in Latin American and the Caribbean? Theory and policy considerations.* Cambridge: National Bureau of Economic Research.

Custodia, N., Wheelock, A., Thumala, D., & Slachevsky, A. (2017). Dementia in Latin America: Epidemiological evidence and implications for public policy. *Frontiers in Aging Neuroscience, 9*, 221.

De Henau, J., Himmelweit, S., & Perrons, D. (2017). *Investing in the care economy*. International Trade Union Confederation. https://www.ituc-csi.org/IMG/pdf/care_economy_2_en_web.pdf. Accessed 23 June 2019.

Haas, H. (2013). Volunteering in retirement migration: Meanings and functions of charitable activities for older British residents in Spain. *Ageing & Society, 33*, 1374–1400.

Hall, K., & Hardill, I. (2016). Retirement migration, the 'other' story: Caring for frail elderly British citizens in Spain. *Ageing & Society, 36*(3), 562–585.

Hawley, C. (2007, August 16). Seniors head south to Mexican nursing homes. *USA Today*. http://usatoday30.usatoday.com/news/nation/2007-08-15-mexnursinghome_N.htm?csp=15. Accessed 23 June 2019.

Houser, A., Fox-Grage, W., & Ujvari, K. (2018). *Across the states: Profiles of long-term services and supports, 2018*. Washington, dc: AARP Public Policy Institute. https://www.aarp.org/content/dam/aarp/ppi/2018/08/across-the-states-profiles-of-long-term-services-and-supports-full-report.pdf. Accessed 23 June 2019.

Kaufer, D. I., Williams, C. S., Braaten, A. J., Gill, K., Zimmerman, S., & Sloane, P. D. (2008). Cognitive screening for dementia and mild cognitive impairment in assisted living: Comparison of 3 tests. *Journal of the American Medical Directors Association, 9*(8), 586–593.

Litwak, E., & Longino, C. F., Jr. (1987). Migration patterns among the elderly: A developmental perspective. *The Gerontologist, 27*(3), 266–272.

McBride, S. (2017, July 1). Ecuador's El Presidente. *New Mobility: the Magazine for Active Wheelchair Users*. http://www.newmobility.com/2017/07/lenin-moreno/ Accessed 23 June 2019.

Norori, M. L. (2018). Addressing the "tsunami" of long-term care needs in Latin America: Is preparation feasible? *Journal of the American Medical Directors Association, 19*(9), 731–732.

Ormond, M., & Toyota, M. (2016). Confronting economic precariousness through international retirement migration. In J. Rickly, K. Hannan, & M. Mostafanezhad (Eds.), *Tourism and leisure mobilities: politics, work, and play* (pp. 134–146). Abington-on-Thames: Routledge Press.

Pot, A. M., Briggs, A. M., & Beard, J. R. (2018). The sustainable development agenda needs to include long-term care. *Journal of the American Medical Directors Association, 19*(9), 725–727.

Sloane, P. D., & Zimmerman, S. (2007, November). *Assisted living in Mexico: An alternative and an opportunity*. Oral presentation at the annual meeting of the Gerontological Society of American; San Francisco, CA.

Sloane, P. D., Cohen, L. W., Haac, B. E., & Zimmerman, S. (2013). Health care experiences of U.S. retirees living in Mexico and Panama: A qualitative study. *BMC Health Services Research, 13*, 411.

Sloane, P. D., Zimmerman, S., & D'Souza, M. F. (2014). What will long-term care be like in 2040? *North Carolina Medical Journal, 75*(5), 326–330.

US Centers for Disease Control and Prevention, National Center for Health Statistics. (2016). *Residential care communities, 2016*. https://www.cdc.gov/nchs/fastats/residential-care-communities.htm. Accessed 23 June 2019.

Warnes, T. A. (2009). International retirement migration. In P. Uhlenberg (Ed.), *International handbook of population aging*. New York/Berlin/Heidelberg/Dordrecht: Springer.

Whelan, J. C. (2012). When the business of nursing was the nursing business: The private duty registry system, 1900-1940. *Online Journal of Issues in Nursing, 17*(2), 6.

# Chapter 8
# Recommendations and Conclusions

**Philip D. Sloane, Sheryl Zimmerman, and Johanna Silbersack**

Twenty-five years ago, anthropologist Leonard Plotnicov proposed that medium-sized cities in Mexico, because of their pleasant climate, charming historic buildings, relative peacefulness, recreational opportunities, and low cost of living, could improve their economies by becoming retirement destinations for older persons from the United States and Canada (Plotnicov 1994). Since then, retirement has become a multi-trillion-dollar industry (McGrath 2018), and retiree-related spending has become an increasingly large component of the economies of many countries, including the United States (Levanon et al. 2018). As we noted in Chap. 1, the actual size of international migration is difficult to judge, but at minimum is in the millions. Still, as of now, Plotnicov's vision is but a partial reality, as immigrant retirees are a minor economic force in Latin America and only in a modest number of its communities.

This book documents how retiree influx has resulted in both economic stimulation and sociocultural change in two Latin American cities that have been particularly successful as international retirement destinations – Cuenca, Ecuador and San Miguel de Allende, Mexico. In this chapter we draw from our interviews and other field work to identify themes and posit recommendations for other communities that may seek to attract international retirees as a means of enhancing their economic development, including strategies for maintaining their identity if they choose to pursue this potential opportunity.

P. D. Sloane (✉) · S. Zimmerman · J. Silbersack
University of North Carolina at Chapel Hill, Chapel Hill, NC, USA
e-mail: philip_sloane@med.unc.edu

© Springer Nature Switzerland AG 2020                                        143
P. D. Sloane et al. (eds.), *Retirement Migration from the U.S. to Latin American Colonial Cities*, International Perspectives on Aging 27,
https://doi.org/10.1007/978-3-030-33543-4_8

## 8.1 Recommendations for Communities Seeking to Attract More Retirees

Our interviews and online surveys identified a number of consistent recommendations that can guide Latin American communities seeking to attract more retirees. The goal is to attract international retirees and to have their presence be a "win-win" situation for both the retirees and the local population. Prominent strategies recommended by our interviewees related to the stability and resources of the country; efforts to improve the community's physical and cultural attractiveness; maintaining and improving environmental quality; attending to community safety; provision of economic incentives and access to health care; making business easy for foreigners to conduct; providing professional development programs for local residents; encouraging collaboration and cultural exchange between the two groups; and actively promoting the community to the outside world. Attending to these areas is likely to help older adults who are natives, as well as help support the community. Each is discussed below.

Be located in a country that is economically and politically stable and has a strong national health care system. National economic and health care policies are not within the control of local governments but are admittedly a key element to attracting retirees. Ecuador, for example, became attractive only as it became economically stable. "*Rafael Correa turned it around*," one interviewee explained, "*by using oil money to build roads, enhance education, and improve health care*," building on the foundation that had been created by adoption of the dollar as currency a few years earlier. Overall economic health in rural areas near each city can also support these destinations by increasing productivity in goods (e.g., fresh produce) and by providing a skilled workforce.

Maximize the community's physical and cultural attractiveness. "*The houses should be freshly painted, and no garbage on the street*", one interviewee explained. Another noted that buildings of historic interest should be restored and maintained. In addition, the community should invest money in activities that would appeal to retirees, as the Instituto Allende and other arts programs did early in the evolution of San Miguel de Allende and continue to do today. The community should also actively encourage and promote cultural events such as festivals, as these attract tourists and engage retirees.

Parks, plazas, museums, and open spaces are important, too. Interviewees pointed out that retirees have more free time, often have dogs that need to be walked, and as a result use parks and public spaces far more than do local residents. An additional recommendation regarding the public areas was a request for ample seating, to be made available for older adults. And, despite the fact that many retirees like to walk, public transportation – such as local buses – is important, and senior discounts for public transportation are appreciated.

Maintain environmental quality. Related to the above point, a key theme was that government must pay attention to environmental quality and not just growth. Park maintenance, preservation of open spaces, and recycling programs were among the examples given. The lack of attention to environment was identified by many inter-viewees as a cause of problems in San Miguel de Allende, particularly around loss of open space and problems with the water supply.

Promote safety. "Security" was a strong theme in both cities. Along with safety comes tranquility, and in that respect the conservative, family-focused tradition of both cities was considered a helpful factor in creating an environment that appeals to many retirees.

Respondents recognized that much of Latin America has a reputation for crime and violence, and that this reputation can be a strong factor dissuading retiree migra-tion. *"They must be away from crime-related routes,"* one Mexican informant said, noting that San Miguel de Allende benefits from not being along narcotic trafficking routes and away from petroleum pipelines, because illegally tapping gas is a major criminal activity in parts of that country.

Security must be tight in retiree areas, and when crime occurs the reaction should be swift. In San Miguel de Allende an expat woman was robbed and killed a few years ago, and the police immediately went to work, found and arrested the culprit, *"and there have not been any more incidents,"* an interviewee said.

Provide economic incentives and access to public health care. Our survey of retirees in the two cities (Table 2.3) indicated that a number of governmental initia-tives to attract international immigration were important to them – most notably exemption from income-related taxes, discounts on utilities and public transporta-tion, and access to public health insurance programs. The survey also indicated that amenity provision such as parks and Spanish classes were rated of equal impor-tance, indicating that factors other than economics are important in attracting inter-national retirees.

Take steps to ease visa, residency, and business processes. Governmental red tape around such activities as obtaining visas, purchasing a residence, or opening a bank account can be frustrating for immigrant retirees; so, countries looking to increase retirement migration should seek to minimize bureaucracy around such activities. Having English-speaking personnel available in city offices is also help-ful in facilitating retiree requests.

Provide public education in business English and elder care. If a city or country were truly interested in capturing some or more of the retirement market, an impor-tant step would be development of educational programs for local residents. As one interviewee explained: *"I would like to see the government prepare better our younger people by providing more English classes, to create more work opportuni-ties...so that when (international retirees) come they will be better prepared, for example, for nursing and care of older persons."*

Such a program of education might include:

- English language training. This training would target persons holding positions in a variety of sectors that would have contact with retirees, such as retail busi-ness, banking, the local police force, transportation, and domestic services;

- Programs to educate and certify for personal care of older persons, particularly home care and nursing skills. Such a program was recently introduced in a high school ("secundario") in San Miguel de Allende;
- Formal settings and opportunities for exchanges ("intercambios") related to language training and cross-cultural friendship development; and
- Concerted efforts to make these opportunities accessible to persons in traditionally low wage positions, such as housekeepers.

Encourage activities and attitudes that bring immigrant retirees and natives together. *"Many (retirees) are obviously rather old and have some difficulty learning the language,"* one local interviewee explained, *"but if we look for ways to bring us together in productive activities it would be better for both groups."* An example given by another respondent was of a "mixed group" of Mexicans and retirees who worked together on a volunteer effort to clean up and enhance a local park in which the playgrounds for children were old and dangerous; the result was an area that was more beautiful and safer for the children. Part of this process involves encouraging the local residents and retirees to be tolerant and respectful of their differences.

Similarly, efforts to inform and educate visitors and immigrant retirees about local history and culture can help with social integration, perhaps as part of language exchanges. In so doing, retirees can also learn about patterns of behavior that are valued in Latin America, such as frequenting the neighborhood tienda and stopping to make small talk with neighbors or store clerks. One tienda clerk said:

> *"For example, the retirees, when they come, they walk their dogs and run into neighbors, and they should chat with them, even if they don't understand the language. It helps you get along because they'll know you as someone who lives in this neighborhood and you get to know as a neighbor, an immigrant... a person who moved here and lives nearby."* Then, remarking about a retiree he had seen in the neighborhood, he added, *"That man, for example, I have never seen here in the store."*

Promote themselves aggressively. Cities should accompany the above efforts with an active, ongoing promotional campaign, possibly that appeals to both tourists and potential retirees. One informant suggested that there may be several distinct promotion models including: (a) a retiree focus, emphasizing a receptive, vibrant, historic community, with inherent cultural resources and little tourism, as has developed in Cuenca; (2) a combination of tourism and retirement, with an emphasis on things to do such as shopping, sightseeing, frequent festivals, and a diversity of restaurants, as has developed in San Miguel de Allende; and (3) a combination of ecotourism and retirement, as is growing in the Guanacaste region of Costa Rica.

Create policies and programs that lead to "win-win" for both natives and foreigners. Most of the local residents we interviewed would agree that growth is better than stagnation or economic decline. Growth must be managed, however, so that it develops in a way that, as one interviewee explained, each party gives something so that everyone can win (*"dar y dar para ganar"*). Such policies require partnerships between public agencies and private enterprise, and, as noted earlier, planning.

A sensitive and challenging issue is whether and to what extent retirees should be taxed to help pay for services and improvements. Several interviewees made this suggestion as a way of helping finance the amenities that will attract retirees. Real estate taxes, for example, are very low in Latin America compared to the United States and could be raised considerably to still be considered quite modest by retirees. On the other hand, low tax rates and exemptions attract retirees; so, the extent to which this option is advantageous is yet to be determined.

## 8.2 Attracting Tourists as a Strategy to Attract Retirees

Tourism is currently a far bigger revenue source for Latin America than is retirement migration. Furthermore, it is often the way by which potential retirees have been first introduced to such popular retirement cities as San Miguel de Allende, Antigua (Guatemala), Los Cabos (Mexico) and Cartagena (Colombia). This connection between tourism and retiree migration is quite understandable, because tourist destinations often possess many of the same amenities that attract retirees, such as an attractive setting, pleasant climate, culture, and restaurants (Hudson et al. 2019). Consequently, it is natural to think of tourism as a key strategic component to attract migrant retirees.

Furthermore, as was noted in Chap. 1, tourism often creates many of the infrastructural elements that make an area fertile for retirement migration. On the other hand, historic colonial cities may already have amenities and cultural resources and therefore not need tourism to create infrastructure – Cuenca being a case in point. So, the importance of tourism as a strategy to attract retirees needs to be considered on a community-by-community basis.

In San Miguel de Allende, much of its ongoing success in retiree growth has been attributed to its popularity among tourists. On the other hand, many of the problems and complaints of local residents and retirees are also due to the hordes of tourists that visit annually. Cuenca, in contrast, lacks prominence as a tourist destination, which helps maintain its cultural identity but may be a factor limiting future growth as a retirement destination. For example, Cuenca lacks an international airport, which means that at least two flights are needed for a retiree to return to the U.S. or Canada, another negative factor for retiree migration.

Therefore, it is a reasonable to question whether tourists and retirees should be attracted simultaneously. Peddicord outlined the factors both for and against (Peddicord 2016):

- Benefits of a tourist town for retirees can include the following components: (1) great weather and, frequently, appealing scenery and landmarks; (2) more and better cafes, restaurants, and entertainment; (3) well-maintained infrastructural elements, such as sidewalks, trails, and beaches; (4) lack of obvious poverty (due to both job promotion and gentrification within tourist areas, with concomitant segregation of poorer persons in other zones); (5) better flight connections; (6) the potential for rental income and resale if the retiree buys a home; (7) more English spoken by local residents in shops, restaurants, and on the street; (8) more first-world conveniences, such as ATMs and familiar franchises; and (9) better construction options if one wants to build.

- Downsides of a tourist town for retirees include: (1) difficulty developing a sense of community, due to feeling that foreigners are transient; (2) greater difficulty integrating with the local culture and its people, because local residents tend to place retirees in the "tourist camp"; (3) higher costs of housing, food, transportation, and other living expenses than non-tourist towns; (4) a feeling that the city is more for show than "real" (the "Disneyland effect"); and (5) tourist annoyances such as noise, traffic, parking problems, beggars, and pickpockets.

Indeed, our research indicates that San Miguel de Allende exemplifies a tourist-retirement town, with both its advantages and downsides, and Cuenca a non-tourist town, with its upsides and downsides.

As Fig. 8.1 illustrates, tourism and retirement tend to stimulate somewhat different sectors of the economy. In that context, from a developmental perspective, tourism and retiree migration can be viewed as complementary.

Tourism development requires investment in infrastructure, in beautification, and in promotion. The best example of this situation is the plan for national tourism development created in the 1960s by the government of Mexico, which was implemented over the subsequent three decades, consumed up to one percent of the national budget annually for many years, and resulted in massive tourism infrastructure development in and around Cancun/Cozumel and Los Cabos (Bringas Rábago 1999).

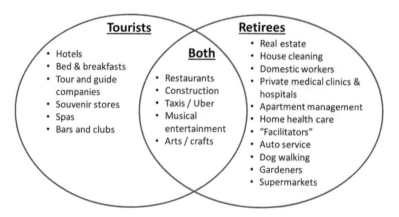

**Fig. 8.1** Similarities and differences between tourists and retirees in business development and job creation

## 8.3    Recommendations for Communities Seeking to Conserve Local Culture and Quality of Life

In many ways, San Miguel de Allende can be viewed as a cautionary tale. Many local residents – and some retirees who have lived there for many years – feel that its success as a tourist and retirement destination is destroying the very atmosphere that created the city's original appeal. When we asked what else could be done to attract immigrant retirees, one local resident responded, *"I believe that we don't want to attract any more; we have more than enough." "Put barriers around the city,"* said another, adding *"My mother said this, that we need to put barriers up…. Because we are going to completely lose the city's identity."*

In a 2017 article, a Mexican anthropologist and former member of San Miguel's tourism board wrote an essay expressing dismay because a hillside that residents had considered a public park was now packed tightly with apartment complexes. The essay included the following passage:

> *"That cultural landscape has changed forever, and a greater damage has been inflicted to it. Now from the Jardín (central square), in front of the Parish, one looks east and instead of looking at the Hill of La Cruz del Pueblo, one looks at a conglomeration of buildings that modify and alter negatively the architectural set of the city, altering the physiognomy that gave the designation Patrimony of the Humanity to San Miguel, causing it to miss an important element that gave it a substantial part of that singularity for which it was inscribed in the list of the UNESCO. (Aveleyra 2017)"*

In this context it is not surprising that, when we asked for suggestions regarding what communities could do to attract more retirees, many respondents from San Miguel did not consider this to be a relevant question. Instead they focused on things that had gone wrong or that needed to be addressed to help other cities maintain or regain their unique "sense of place" in the face of popularity as a retiree and/or tourist destination.

Among the themes expressed around preservation of local culture were governmental reform, measures to reduce economic disparities and increase services for all community members, involvement of retirees in community development, taking steps to minimize congestion in the central city, championing open spaces, and encouraging development on the outskirts rather than in the city center. Each is discussed briefly below.

Promote governmental accountability. A subtle theme in our interviews was that governmental authorities were sometimes championing growth without considering the welfare of the city's overall population. *"I smelled a lot of corruption and lack of care for the public welfare,"* one of our data collectors (a social worker) noted in her field notes after interviewing a government worker. *"The most telling answer by the director is that municipal expansions and improvements don't happen based on need, but demand,"* she wrote. This sentiment was echoed by interviewees as well. *"Eliminate governmental corruption,"* one said, adding *"and have less bureaucracy.*

*Because I believe that bureaucracy generates corruption.*" Such issues are delicate, as governments are easy to blame but hard to effectively replace, and both of our study cities have elite classes and families that are highly involved in both government and business enterprises within the community.

Reduce disparities and increase services for all community members. Interviewees pointed out that both cities continue to have significant portions of their populations living at or below the poverty level, including many older persons. The gap ("la brecha") between the wealthy (including most retired foreigners) and the service workers in the community has become both greater and easier to ignore, as gentrification has led to de facto segregation by income between the wealthier central and the less wealthy peripheral communities.

Potential measures to ameliorate this situation include taxes that would differentially affect the wealthy (e.g., on real estate) and could then be used to promote public education in skills that would provide upward mobility for workers. Additionally, governments should take care to provide equivalent and high-quality public services, such as water, trash collection, and police protection, in all neighborhoods to an equal degree. Immigrant retirees could play a role as well, for example by using local tiendas more and chain supermarkets less. Central to such activities would need to be a development plan to reduce the gap between rich and poor.

Involve retirees in community development and events. As noted elsewhere in this book, volunteering and community service by immigrant retirees are common in both Cuenca in San Miguel de Allende. Increasing these efforts and channeling them in a thoughtful way toward preserving community identity was suggested by some of our interviewees. *"The number one thing is a little more tolerance on the part of the Mexicans and a bit more politeness on the part of the foreigners,"* said one interviewee, adding, *"Politeness in the sense of being a good neighbor."* For retirees who do not know Spanish well, participation in park and open space beautification and preservation might be a method of channeling the volunteer spirit toward community development. For all retirees, efforts to better educate and involve them in common community events could help them become participants rather than only spectators.

Minimize traffic congestion in the central city. Both San Miguel and Cuenca have traffic congestion, in large measure because the narrow streets of the old parts of the city become easily congested with motor vehicles. Measures that have been proposed to ameliorate this situation include closing some streets to motor vehicles, particularly on weekends; providing free buses (ideally electric, so as to be non-polluting) between the city center and selected outlying areas, where would be located municipal parking facilities; limiting commercial vehicle access to early mornings; and aggressively towing illegally parked vehicles.

Maintain and preserve open spaces. Parks and open spaces have not traditionally been generously allocated in Latin American cities. However, as long as land prices were modest and the cities had not grown too large, open spaces remained – sometimes because some land parcels were difficult to build on and therefore became de facto community parks. That situation changes with growth, though, whereby vacant land disappears and with it the potential to preserve the green spaces that residents value. Therefore, an important component of community preservation during periods of growth is to specifically put resources into open space preservation and the development and maintenance of parks, which not only serve as an attraction for retirees but also as valuable assets for all community members.

As a case in point, residents in Cuenca take pride in the fact that their city has more parks than many other cities. As noted in Chap. 1, the parks resulted in part because the city was founded at the confluence of several rivers, meaning that building too close to the water would invite flooding. As the city grows, its leaders must also look for areas in the periphery that have not yet been developed and can be preserved for future parks.

Encourage development on the outskirts rather then in the city center. One interviewee noted that a retirement village largely for expats entitled *los Labradores* had been developed several kilometers outside of San Miguel de Allende. As detailed in Chap. 7, *Los Labradores* is very much like a continuing care retirement community (CCRC) in the states (AARP 2011), providing residents with gated community security, comprehensive services, recreation, and health care. It also provides its residents with ready access to the city center by way of a free, regular shuttle service, thus offering the amenities of city retirement without adding to overcrowding.

## 8.4   A Matter of Balance

Could a national development strategy around attracting international retirees create an economic engine for a country such as Mexico or Ecuador? The success of Mexico's tourism development initiative of the 1960s, 1970s, and 1980s suggests that the answer may be "yes." Add to this the fact that the U.S. baby boomer generation includes 75 million people, has a history of adventurousness and tradition-breaking, has retirement resources that are often modest by U.S. standards but that represent significant wealth on a world-wide scale, and faces rising out-of-pocket health and long-term care costs – and a major boom in retirement migration feels realistic.

On the other hand, retirees are especially skittish about safety issues (Teh Cheng Guan 2018). A recent example is the violence that erupted in Nicaragua in the spring of 2018, which significantly slowed the growth of international migration to Granada, another city that we had considered including in our research. Other examples are the rise of drug and gang-related violence in Mexico and Central America and the rapidity with which Venezuela devolved into a chaotic, violent, economically failing country. As one retiree in Cuenca told us (Chap. 3), "*In Latin America, you are only one president away from a Venezuela.*"

There are other barriers as well. An important one is the key role that proximity to family plays in retirement decisions (Hudson et al. 2019); which less distant communities, including rural areas closer to home, can parlay into initiatives to attract retirees (Stallmann and Siegel 1995). Health care and health insurance are other major issues, because North American retirement migration lacks the reciprocity of public health insurance that has made international retirement much more common and attractive in Europe. Also, differences in language and culture, while attractive to some, are barriers to many.

What, then, is the future for international retirement migration, and particularly of North Americans to Latin America? As demonstrated through our research with retirees and a range of local natives, a matter of "balance" may indeed be the operative issue.

- There are advantages and disadvantages for the retirees, and in the end, the balance for them is one that is personal, and so boosting retirement migration will require tipping the balance so that the real advantages of relocating to another country outweigh the real disadvantages.
- There are advantages and disadvantages for potential retirement destinations as well, and here the communities must consider what their vision of "success" looks like and whether, on balance, the opportunity is there, the resources are worth expending, and potential negative consequences can be minimized.

Through the time we spent in discussions with local interviewees, retirees, and experts in the field, it became evident that retiree migration can be a positive or negative phenomenon within a community, and that whether it will catch on more widely remains in the balance. What is clear is that an optimally positive outcome for both retirees and local populations will require careful consideration and planning, so that development does not come at the expense of the local populations' wishes and needs.

# References

AARP. (2011). *About continuing care retirement communities.* https://www.aarp.org/caregiving/basics/info-2017/continuing-care-retirement-communities.html. Accessed 29 July 2019.

Aveleyra, A. (2017, October). San Miguel de Allende, Point of No Return. *Lokkal Magazine San Miguel de Allende.* https://www.lokkal.com/sma/magazine/2017/october/alberto.php. Accessed 29 July 2019.

Bringas Rábago, N. L. (1999). Politicas de desarrollo turístico en do zonas del pacifico mexicano. *Región y Sociedad, 11*(17), 3–51.

Hudson, S., Fung So, K. K., Li, J., & Cárdenas, D. (2019). Persuading tourists to stay – Forever! A destination marketing perspective. *Journal of Destination Marketing & Management, 23,* 105–113.

Levanon, G., Anderson, B., Cheng, B., & Steemers, F. (2018). *The impact of demographic trends on US Consumer Spending.* New York: The Conference Board.

McGrath, C. (2018, June 21). U.S. retirement assets at $28 trillion in Q1, little changed from end of 2017. *Pensions & Investments.* https://www.pionline.com/article/20180621/INTERACTIVE/180629958/u-s-retirement-assets-at-28-trillion-in-q1-little-changed-from-end-of-2017. Accessed 29 July 2019.

Peddicord, K. (2016, February 10). The pros and cons of retiring in a tourist town. *US News Money*.            https://money.usnews.com/money/blogs/on-retirement/articles/2016-02-10/the-pros-and-cons-of-retiring-in-a-tourist-town

Plotnicov, L. (1994). El atractivo de las ciudades medias. *Estudios Demográficos y Urbanos, 9*(2), 283–301.

Stallmann, J. I., & Siegel, P. B. (1995). Attracting retirees as an economic development strategy: Looking into the future. *Economic Development Quarterly, 9*(4), 372–382.

Teh Cheng Guan, B. (2018). Retirement migration: The Malaysia My Second Home (MM2H) program and the Japanese Retirees in Penang. *International Journal of Asia-Pacific Studies, 14*(1), 79–106.

# Appendices

## Appendix A: Overview of the Study and Its Research Methods

The goals of this research study were fourfold: (1) to investigate the impact of retired foreigners on colonial cities, (2) to investigate the opinions of local residents and retired foreigners on retirement migration, (3) to understand the role of local and national government, and (4) based on this information, to provide guidance on how cities may manage the influx of migration in a way that supports the local community and growth of the economy, while preserving local culture and minimizing potential negative impact.

To provide a comprehensive picture of these issues from multiple viewpoints, the study employed a variety of data collection strategies: in-person, semi-structured interviews of local community members stratified into certain professional groups; an online survey of retired immigrants from the United States, Canada, or Europe; in-person interviews of retirees active in the community; semi-structured interviews of retired immigrants over the age of 70 who use long term care services within the city of interest; field notes by each member of the study team; geographic mapping of retiree and local interviewee residences, including creation of density maps; photographs; and, as available and relevant to our research questions, demographic data from governmental sources.

### Funding

Primary funding was provided by Research Grant # HJ-131R-17 from the U.S. National Geographic Society, entitled Retirement Migration from the U.S. to Latin American Colonial Cities: Implications on Individual and Community Health,

© Springer Nature Switzerland AG 2020
P. D. Sloane et al. (eds.), *Retirement Migration from the U.S. to Latin American Colonial Cities*, International Perspectives on Aging 27,
https://doi.org/10.1007/978-3-030-33543-4

Conservation of Local Culture, and Quality of Life. Secondary support was provided by the Elizabeth and Oscar Goodwin Endowment of the Department of Family Medicine at UNC.

## Setting

Two cities were studied: Cuenca, Ecuador and San Miguel de Allende, Mexico. Both are historic colonial cities and UNESCO World Heritage sites, with deep cultural roots. Both are also popular international retirement destinations. Additionally the two cities were believed to be at different stages in development as retirement destinations, with San Miguel de Allende having been popular for several decades, whereas Cuenca only in the last 10 years, and with San Miguel de Allende being rather small (less than 150,000 population) and Cuenca being larger (over 500,000) and economically more diverse. Further details about the two cities are presented in Chap. 1.

## Study Team

Drs. Philip Sloane and Sheryl Zimmerman, professors at the University of North Carolina at Chapel Hill, co-directed the project, with Johanna Silbersack, MSW, acting as the project coordinator.

Data collection was conducted primarily by a team of four students and recent graduates from UNC. The research assistants were trained during 2 weeks, which was followed by 3 weeks of data collection in each city and a week reviewing the data and developing a coding scheme. In the subsequent 3 months a team of three UNC graduate students, all fluent in Spanish, coded the data, which were then analyzed by the research team.

In each of the two cities, the study was aided by a local coordinator with expertise in research (one was a doctoral student, the other a retired engineer), with whom the principal investigator met personally several months before the study to review the methods, and who (a) in the weeks before data collection identified potential interviewees and arranged interview appointments, (b) met with the study team on the first day to brief them on both the schedule for the first days and aspects of conducting research in the local culture, and (c) accompanied each research team member on several initial interviews.

## Protection of the Rights of Human Subjects

All research methods followed conventions regarding informed consent, confidentiality of data, and preservation of anonymity of respondents. Verbal, embedded and/ or written consent was obtained from all participants of this study. All study methods were reviewed and approved by the Institutional Review Board of the University of North Carolina at Chapel Hill (#18-0427).

## Semi-structured Interviews of Local Residents in the Study Cities

Interviewing local residents within each city were the primary method of data collection, due to the gap in literature representing local communities' opinions and perspectives on international retirement migration. In each city, our bilingual research team sought to complete a minimum of six interviews with local residents in each of six categories (36 interviews in each city), chosen to represent a diversity of opinions regarding the impact of retirement migration largely among individuals likely to encounter and know about immigrant retirees.

We identified potential participants using a combination of networking by local coordinators, internet searching, word of mouth, and (in the case of the tiendas) stratifying neighborhoods into either high or low retiree concentrations and touring the neighborhoods for potential participants. The research team obtained verbal consent in Spanish from each participant and provided a written consent addendum for participants who permitted their interview to be audio-recorded or consented to video and/or photography.

The interview categories and their recruitment methods are briefly summarized below:

- Real Estate Agents. They were identified by the local coordinators within each city, by word of mouth, and through internet searches conducted before data collection. Eligibility criteria included: native of the country in which they were being interviewed and working for a real estate company that had an English advertisement on the internet. All 12 interviews (6 in each city) were conducted in Spanish. The interview protocol (Appendices B and C) included an addendum of questions regarding the real estate market.
- Government Officials. The majority of government officials interviewed were identified and arranged by the local coordinators within each city. In each city, we attempted to interview a range of individuals, including persons from the mayor's office, municipal public services, offices of tourism and foreign affairs, and planning departments. All 17 interviews (9 in Cuenca, 8 in San Miguel de Allende) were conducted in Spanish. The interview protocol (Appendices B and C) included an addendum that asked additional questions related to strategic plans regarding immigration.
- Health Care Workers. Health care workers were identified via internet searching before data collection occurred, by the local coordinators, and by word-of-mouth within each city. Interviewees were chosen to represent a range of settings and perspective, with a focus on individuals who had contact with immigrant retirees, including hospital administrative staff, physicians, dentists, and home care nurses. All 13 interviews (6 in Cuenca, 7 in San Miguel de Allende) were conducted in Spanish.

- Non-governmental services providers. Service providers based in each community were identified via word of mouth, networking throughout the local and expatriate communities, and by local coordinators. The term "service provider" was defined as a person who offered some service (whether non-profit or for-profit) to the community at large. This allowed the research team to interview a range of persons that represented the interests and needs specific to Cuenca and San Miguel's communities, including directors of local nonprofits service agencies, administrators of services for older adults, local, well-established businesses and cafes, and facilitators (i.e. persons who specifically market their trade towards retired expats). All 13 interviews (6 in Cuenca, 7 in San Miguel de Allende) were conducted in Spanish.
- Workers in Small Neighborhood Convenience Stores ("tiendas"). This group was recruited in two strata by location – low-retiree density neighborhoods and high-retiree density neighborhoods – and so our goal was to conduct 12 interviews (6 per stratum) in each city. These interview categories allowed the research team to speak with a standardized subset of the local population that is highly connect to local communities. To identify tiendas to interview, we used data from our real estate interviews to create density maps each city based on retiree concentration (see Chap. 3 for details on this method). Thus, the tiendas were identified and the interviews conducted during the second half of the data collection period in each city. All 24 interviews (12 in each city) were conducted in Spanish. Interviews with tienda employees included an addendum that asked additional questions related to pricing and availability of items within their shops (see Appendices B and C for copies of the interview instruments).

The interview protocol (Appendices B and C) included a combination of open-ended questions and Likert-scale questions. The interview was designed to last approximately 45 min; actual interview length varied between 20 and 90 min. All participants provided verbal informed consent prior to beginning the interview, and separate, written consent was obtained for audio recording. All interviews included three topics, each of which included multiple open-ended questions, often with probes to elicit specific information:

- effect of retiree immigration on the city in general, on quality of life of local citizens, on the environment, and on public safety and security;
- social relationships between the immigrant retirees and the local population, including communication, customs (including cultural differences), patterns of socializing, and comparison between immigrant retirees and tourists; and
- government support of and reaction to the immigrant retirees.

As noted above, supplemental questions were asked of real estate, governmental and tienda respondents.

**Online Survey of Immigrant Retirees (Aged 55 and Older) in the Study Cities**

An online survey was the primary method by which the study team gathered data on the community of immigrant retirees living in our two cities of interest. This method was chosen due to its efficiency, ability to reach large numbers of retirees, and to ensure that data collection time was not taken away from the in-person interviews, which were the primary focus of the research.

Recruiting was primarily done online and was felt to be appropriate, since nearly all members of the immigrant retiree community are online regularly, and many communicate through blogs and Facebook pages. To promote the survey in each city, our research team networked online, by email, and in person with leaders in each expat community who had an online presence and posted notices on online discussion forums, yahoo groups, news sites, and wherever else we were able to obtain permission to announce the survey. In addition, the research team disseminated flyers with online survey information in coffee shops, restaurants, and other locations frequented by expats.

The online survey, designed to take approximately 20 min to complete (refer to Appendix E), included a mix of multiple choice, open-ended and Likert-scale questions. All respondents, before being given access to the survey, were displayed an embedded consent form. Respondents were required to answer three screening questions before they were able to move forward in the survey as well; these were to confirm that all respondents were over the age of 55, had moved to Cuenca or San Miguel de Allende after the age of 50, and had spent a minimum number of months as a resident of the city. The primary topics of the survey included demographic characteristics of retirees, retirees' involvement in the local community, retirees' impressions of the government benefits in each city, and open-ended probes on the theme of retirement migration.

**Interview Form for Immigrant Retirees Using Long-Term Care Services**

An additional research question of the team surrounded the availability and types of long-term care services, since older persons often reach a point where these are needed. To systematically gather data on individuals using such services, we designed an in-person supplement to the online survey to be given to retirees over the age of 70 in each city, who currently use either home-based or residential long-term care services. We anticipated having difficulty finding and interviewing such individuals, and so the study goal was to obtain a minimum of three interviews in each city.

To be eligible, interviewees had to be aged 70 or older, be a retired immigrant to the city of study, have immigrated after the age of 55 to the study country, and either use formal (i.e. paid) home long-term care services daily or live in a residential long-term care setting. Long-term care services were defined as services related to bathing, dressing, and/or medication use, whether in a personal home or in a facility. Persons were identified via word of mouth.

The research team publicized their interest in speaking with long-term care users while networking with retired immigrants and, as per the study's IRB protocol, asked to be contacted if such an individual was interested in being interviewed. Once an eligible participant reached out to the research team, an in-person interview was arranged. Verbal consent was required for the interview and written consent for audio or video recording, or for photographs.

A copy of the interview form is included as Appendix F.

## Data Collection

We conducted data collection in Cuenca in June 2018 and in San Miguel de Allende, Mexico during July 2018. Interviews were conducted by four trained bilingual research assistants, who traveled as a team to each of the study sites, with support from a local coordinator with expertise in the conduct of research, and with the principal investigator on site to assist with selection of interview subjects, adherence to project protocols, and general project management. Additionally, in each study city, a city-specific online survey of immigrant retirees was fielded during week two of data collection – at a time when the study team was in the city and had already met and presented the city to local retiree leaders and could further promote it using flyers and interviews.

Before their visits to both locations, the local coordinators and the team had made prior contact within each city to begin the process of setting up interviews with local residents as well as networking with retired immigrants, and the project manager contacted leaders among the expat communities in each city who were active in online blogs or retiree organizations.

By the end of the data collection period, the team had exceeded the goal of 36 interviews within each city, conducting 39 in Cuenca and 40 interviews in San Miguel de Allende. Additionally, 5 in San Miguel and 2 in Cuenca were conducted in English, with non-local representatives from the a few of the six categories, including real estate, health care, and service providers. The majority of these interviews were audio recorded (84%) and transcribed into Spanish using an online program called Trint (Trint LTD, London, England); when interviewees desired not to be recorded (16%) the interviewers took careful notes and formally wrote them up. Each transcription was reviewed and edited by the data collector who conducted the interview and supplemented by written notes taken during each interview.

We also completed our desired three interviews of retirees using long-term care services in each community, plus a variety of other interviews with retirees and local residents, visits to retiree homes and gathering places, informal meetings with retirees and local residents, and visits to volunteer and service sites in which retirees were involved. The online surveys remained active for 4 weeks after the on-site data collection, gathering a total of 424 completed surveys from Cuenca and 325 from San Miguel de Allende.

**Photographs and Videos**

In the process of carrying out our field work, the research team gathered photographs and video recordings of the general setting of each city, of interviewees who consented to be photographed and/or recorded (through the written consent process referenced above), and of homes and convenience stores ("tiendas"), since real estate and tiendas were areas of focus for the study. Throughout the course of the data collection, the research team collected thousands of photographs, some of which are included in this book.

**Mapping**

An additional element of the on-site data collection was use of mapping techniques to document the distribution of immigrant retirees in both cities. This involved asking realtors and other interviewees in each city to indicate on a map where expat retirees live, by marking each neighborhood with a 0 (none), 1 (few), 2 (many) and 3 (most). Altogether 18 interviewees in San Miguel de Allende and 10 in Cuenca felt they could reliably identify on each map these areas and therefore contributed data to the maps.

We combined the data from our informants in each city (18 in San Miguel de Allende and 10 in Cuenca) in order to create density cloud maps of both cities, which illustrated the prevalence of retirees by neighborhood. This was supplemented by points of location given by the expatriate retirees (see below for further detail). This multi-method approach was inspired by Álvarez et al. (2017) who had effectively used geo-mapping in a recent study conducted by the municipality of Cuenca, in which they gathered mapping data from four different sources and found that they all generated similar results; so we chose the method that would generate data most quickly (interviews of local residents), as these data were needed for identifying where to conduct our tienda interviews, and then supplementing it with data from the online survey of retirees.

To further identify and display the distribution of residents in the city, we invited each local resident we interviewed and each retiree who completed our online survey to place a dot on a map indicating where they lived. All told, 36 of our 39 local resident interviewees in Cuenca and all 40 of our local interviewees in San Miguel de Allende provided data. Our response rates to the online survey were less complete – 300 of our 424 respondents in Cuenca and 210 and 325 in San Miguel de Allende.

## Data Analyses

Coding of transcribed interviews with local residents. The entire research team reviewed and discussed all transcripts, and word-by-word reading was used to derive codes – such as relationships, communication, culture, and characteristics of retirees. Then codes were sorted into categories (e.g., social dynamics, real estate, recommendations) which included sub-codes (e.g., harmony, conflict, recommendations for integration). The advantage of this approach is that preconceived categories were not imposed upon the data. The research team ultimately settled on 59 unique codes that were grouped into 17 overarching categories. An additional three categories/codes were used in order to identify quotes, stories, as well as a code to identify unclear themes from the data. Appendix D contains the final codebook, including definitions for each code.

After a final codebook was developed, an additional three, fluent, Spanish speakers not involved in data collection were trained in qualitative coding. During training, they independently coded interviews to compare and contrast their agreement per code. After this process was repeated for seven interviews, the research team achieved enough agreement per code to independently code the remaining interviews.

Analyses of interview transcripts and field notes. Analyses used NVivo (https://www.qsrinternational.com/nvivo/home) and followed principles of conventional content analysis, (Hsieh and Shannon 2005) and inductive category development (Mayring 2000). The entire research team reviewed and discussed all transcripts, and word-by-word reading was used to derive codes – such as relationships, communication, culture, and characteristics of retirees. Then, codes were sorted into categories (e.g., social dynamics – which included relationships, harmony, conflict, segregation, integration, and other codes), which ultimately identified key themes. The advantage of this approach is that preconceived categories were not imposed upon the data.

Codes from the local resident interviews (Appendix D) that were analyzed for Chap. 4 included relationships, harmony, conflict, segregation, integration, paid relationships, power/privilege, philanthropy, communication, communication barriers, cultural similarities and differences, positive and negative characteristics of retirees, prejudice, attitudes about internal migrants, migrants returning to their home country, and international migration from other Latin American countries. Codes analyzed for Chap. 5 included relationships, paid relationships, power and privilege, the current economy and the economy as related to retirees; codes analyzed for Chap. 6 included all health care codes – including attitudes regarding retirees' use of health care, cost of health care, current state of health care, and the quality of health care (with a focus on health care provider interviews); and codes analyzed for Chap. 8 included those related to recommendations (recommendations, potential policies, recommendations for integration, and deter expat migration).

Analyses of quantitative data from the interviews and online surveys. These consisted of continuous data (such as age) and categorical data. Traditional basic data analysis techniques were used, including the computation of descriptive statistics and testing for associations using appropriate statistical methods for the type of data being analyzed. For simple computations we used Excel; for more complex computations we used software from the Statistical Analysis System (SAS – https://www.sas.com/en_us/home.html). Descriptive statistics were used to describe the sample overall, as well investigating differences in respondents by city; results within this book are largely reported in means, percentages, and standard deviations. Other bivariate analysis included Spearman and Pearson chi-squares, Cochran-Mantel-Haenszel ordered chi-square, and independent sample t-tests.

## References

1. Álvarez, M. G., Guerrero, P. O., & Herrera, L. P. (2017). *Estudio sobre los impactos socio-económicos en Cuenca de la migración residencial de norteamericanos y europeos: aportes para una convivencia armónica loca. Final Report.* Cuenca: Avance Consultora.
2. Hsieh, H. F., & Shannon, S. E. (2005). Three approaches to qualitative content analysis. *Qualitative Health Research, 15*(9), 1277–1288.
3. Mayring, P. (2000). Qualitative content analysis. *Forum: Qualitative Social Research, 1*(2). Retrieved June 30, 2019, from http://www.qualitative-research.net/fqs-texte/2-00/02-00mayring-e.htm

# Appendix B: Interview Form for Local Residents – Spanish Version

*Quisiera preguntarle cómo siente usted que la inmigración de jubilados extranjeros está afectando o puede afectar [ciudad], incluyendo el área alrededor de la misma.*

1. ¿En su opinión, es bueno para la comunidad que los inmigrantes jubilados se muden aquí? ¿De que manera?

> Sondeo: En su opinión, ¿Hay inconvenientes causados por esta inmigración y, si es así, ¿cuáles son?

2. De acuerdo a su conocimiento, ¿cuáles han sido los efectos tanto positivos como negativos en la calidad de vida de los ciudadanos locales?

> Áreas para sondear: trabajos / mercado de vivienda y precios / transporte / carácter de los vecindarios / actitudes actividades de los jóvenes / actitudes tradicionales.

3. A continuación, hablemos sobre el medioambiente y los problemas del desarrollo sustentable. Me gustaría saber qué temas son importantes para mantener la calidad ambiental si la población continúa creciendo.

> Áreas para sondear: agua / aguas residuales / transporte / áreas públicas de recreación como parques?

4. A continuación, me gustaría hablar sobre temas de seguridad pública. En su opinión, ¿el movimiento de jubilados extranjeros en la [CIUDAD] tuvo un impacto en la seguridad de todas las personas que viven aquí?

> Sondeo: ¿La ciudad tuvo que modificar o aumentar los servicios de seguridad debido a la presencia de extranjeros?

5. Una última pregunta en esta sección. Se trata de las diferencias económicas dentro de la población nativa. ¿Hay formas en que el impacto de la migración de jubilados afecte a algunos, pero no a otros en la comunidad?

> Sondeo: Por ejemplo, tal vez los efectos negativos en el medio ambiente puedan ser sentidos por todos, pero los beneficios económicos sólo por algunos.

## Parte 2

*Ahora pasamos a la segunda parte de la entrevista.*

1. ¿Cómo describiría en general la relación entre los ciudadanos locales y los inmigrantes jubilados? ¿Con esto me refiero a los tipos de contacto entre ellos, y el tipo de relaciones que se desarrolla?

2. A veces surgen conflictos entre las costumbres locales y las actividades de los inmigrantes jubilados. Cuéntenos acerca de una experiencia que haya tenido sobre dicho conflicto y qué recomendaría que las comunidades puedan hacer para prevenir o minimizar dichos conflictos

3. Algunas personas piensan que los inmigrantes jubilados son una especie de turista. Cuando usted piensa en los beneficios y desventajas de la comunidad, en su opinión, ¿cómo se <u>comparan los jubilados inmigrantes con los turistas más tradicionales</u>, como los estudiantes o las personas mayores, que vienen por un tiempo breve y luego se van a casa?

4. ¿Cuáles son sus impresiones sobre la integración de los inmigrantes jubilados aquí?

> Sondeo: ¿Hay diferentes subgrupos o tipos de inmigrantes jubilados?
> Sondeo: Si la mayoría de los inmigrantes jubilados viven separados, ¿Es esto un problema o no?
> Sondeo: En su opinión, ¿qué tipo de acciones y actividades pueden unir mejor a las dos comunidades?

5. ¿Hay otras cosas que quisiera decir sobre la presencia de jubilados inmigrantes en [CIUDAD]?

*Si el entrevistado es un agente de bienes raíces, un trabajador del gobierno o dueño de una tienda de conveniencia/ pulpería / venta / recaudería / abarrote, haga ahora las preguntas relevantes de las últimas 2 páginas de este formulario*

## Parte 3

*Ahora pasamos a la tercera parte de esta entrevista.*

Algunos países o ciudades de América Latina tienen leyes diseñadas para atraer a los inmigrantes jubilados. Algunos ejemplos incluyen la ayuda en ciertos impuestos y descuentos en ciertos servicios.

1. ¿Qué beneficios conoce usted que actualmente ofrecen en [PAÍS] y en [CIUDAD] a los inmigrantes jubilados? ¿Cómo se siente usted acerca de darles esos beneficios?


2. Uno de los beneficios que los inmigrantes retirados a menudo desean es el acceso a la atención médica local. ¿Cuál es su opinión acerca de si los inmigrantes retirados deberían inscribirse en los servicios públicos de salud y en que circunstancias?


3. Si su comunidad quisiera atraer más jubilados inmigrantes aquí, ¿En su opinión qué cosas adicionales podría hacer el gobierno que pudiera ser más útil?

## Parte 4

*[Presente la tarjeta laminada con el acuerdo de las opciones de respuesta]*

*A continuación, mencionaré algunas afirmaciones. Por favor responda a cada una diciendo si está completamente de acuerdo (4), de acuerdo (3), ni de acuerdo ni en desacuerdo (2), en desacuerdo (1) o completamente en desacuerdo (0). Recuerde que le pedimos su propia opinión como individuo y no como representante de su trabajo. Además, sus respuestas serán completamente confidenciales y serán combinadas con muchas otras para entender mejor cómo se siente la comunidad con respecto a estos problemas.*

|  | 0 = Completeamente en desacuerdo / 1 = en desacuerdo / 2 = ni de acuerdo ni en desacuerdo / 3 = de acuerdo / 4 = completamente de acuerdo |
|---|---|
| 1. Tener inmigrantes retirados en mi ciudad es bueno porque crea empleos. | 0   1   2   3   4 |
| 2. Las fabricas, fincas, y otros negocios locales no vinculados al turismo han sufrido a causa de la inmigración de los jubilados extranjeros. | 0   1   2   3   4 |
| 3. Los precios de la propiedad inmobiliaria han aumentado debido a la demanda por parte de los inmigrantes jubilados. | 0   1   2   3   4 |
| 4. Los agentes de bienes raíces prefieren vender o alquilar a los inmigrantes jubilados. | 0   1   2   3   4 |
| 5. Los extranjeros están causando que los residentes locales se muden del centro a las afueras de la ciudad. | 0   1   2   3   4 |
| 6. Los residentes locales que desean tener éxito en los negocios necesitan aprender inglés. | 0   1   2   3   4 |
| 7. Tener a Donald Trump como presidente de los Estados Unidos no ha sido bueno para las actitudes locales hacia los Estados Unidos y los estadounidenses jubilados. | 0   1   2   3   4 |
| 8. La inmigración de jubilados extranjeros aquí ha conllevado a una mayor necesidad de servicios policiales y de seguridad. | 0   1   2   3   4 |
| 9. Tener muchos extranjeros jubilados en la ciudad es malo para la calidad de vida de los nativos. | 0   1   2   3   4 |
| 10. Los jubilados inmigrantes son mejores para la comunidad que los turistas. | 0   1   2   3   4 |
| 11. La presencia de tantos extranjeros ha cambiado la cultura de nuestra ciudad. | 0   1   2   3   4 |
| 12. Partes de nuestra ciudad son una colonia norteamericana. | 0   1   2   3   4 |
| 13. Las personas que se mudan aquí desde otros países deberían aprender español. | 0   1   2   3   4 |
| 14. Los inmigrantes jubilados aumentan el costo de vida para los locales. | 0   1   2   3   4 |

| | |
|---|---|
| 15. A menudo los jubilados inmigrantes no respetan la cultura local. | 0  1  2  3  4 |
| 16. Los norteamericanos no entienden la forma en que nosotros hacemos las cosas aquí. | 0  1  2  3  4 |
| 17. La presencia de inmigrantes jubilados me hace sentir como un ciudadano de segunda clase en mi propia ciudad | 0  1  2  3  4 |
| 18. La presencia de inmigrantes jubilados enriquece a los ricos, pero no ayuda a los pobres | 0  1  2  3  4 |
| 19. Los inmigrantes jubilados hacen obras de caridad y trabajo voluntario que es valorado por mi comunidad | 0  1  2  3  4 |
| 20. Los jubilados inmigrantes actúan de manera privilegiada cuando están en espacios públicos | 0  1  2  3  4 |
| 21. Los inmigrantes jubilados son amistosos y abiertos en relación con los nativos de nuestra ciudad. | 0  1  2  3  4 |
| 22. Los inmigrantes jubilados son una mala influencia para los jóvenes en nuestra ciudad. | 0  1  2  3  4 |
| 23. Me gustaría ver que más inmigrantes jubilados se muden aquí para vivir. | 0  1  2  3  4 |

*Casi terminamos. Tengo dos preguntas sobre usted y luego una última oportunidad para que diga lo que piensa.*

1. ¿Cuánto tiempo lleva viviendo en [CIUDAD]?                              _____ años.

2. ¿En qué parte de la ciudad vive actualmente? [Muestre un mapa de la ciudad y pídale a la persona que ponga un punto en ese mapa]

3. ¿Cuál es su ocupación?                 _____

4. ¿Cuántos años tiene usted?                              _____ años.

5. ¿Desde su punto de vista hay algo más que le gustaría decir sobre todo lo que hemos hablado?

## Preguntas para personas que trabajan en el gobierno

1. ¿Qué porcentaje de los ingresos de la ciudad proviene de impuestos a la propiedad?

_____ %

2. ¿Existe un plan estratégico local para abordar la inmigración? Por favor descríbalo.

Sondeo:  Si hay una copia, ¿pudiera facilitarnos"

3. ¿Hay programas que se hayan desarrollado o se hayan instituido cambios en el presupuesto en la ciudad para atender mejor a los inmigrantes jubilados?  De ser así, explique por favor.

4.  Algunas personas sugieren que el gobierno de la ciudad debería cobrar más impuestos a los inmigrantes jubilados, como impuestos sobre bienes raíces, para pagar mejores servicios para ayudar a toda la comunidad, especialmente a las personas de las áreas más pobres. ¿ Usted apoya esta idea? De ser así, ¿cuáles considera que son las prioridades de gasto?

Sondeo: ¿Cuáles son las áreas de mayor necesidad comunitaria en términos de servicios gubernamentales?

5. La falta de oportunidades de trabajo para los hombres a menudo es un problema en las comunidades que dependen del turismo, incluidas las comunidades con grandes poblaciones de jubilados. ¿Es este un problema en [CIUDAD]? y, de ser así, ¿cuáles podrían ser las soluciones que ayudarían a mejorar la situación?

6. ¿Cómo se podría beneficiar más el gobierno de [CIUDAD] de la inmigración de jubilados de América del Norte?

## Preguntas para personas que trabajan en bienes raíces

1. Aquí hay un mapa de la ciudad dividido en zonas. Para cada zona, me gustaría que califique, utilizando su conocimiento de la propiedad inmobiliaria, la cantidad de inmigrantes jubilados que viven ahí en una escala de 0 a 3, donde 0 = ninguno; 1 = pocos; 2 = algunos; 3 = muchos.

*En las dos siguientes preguntas hablamos de compras de viviendas de la ciudad:*

2. Aproximadamente, ¿qué porcentaje de las nuevas compras de bienes raíces en el centro histórico son realizadas por inmigrantes jubilados?

_____ %

3. Aproximadamente, ¿qué porcentaje de las compras de bienes raíces en toda la ciudad corresponden a inmigrantes jubilados?

_____ %

*En las dos siguientes preguntas hablamos de viviendas de alquiler:*

4. Aproximadamente, ¿qué porcentaje de las nuevas rentas de bienes raíces en el centro histórico de la ciudad son realizadas por inmigrantes jubilados?

_____ %

5. Aproximadamente, ¿qué porcentaje de las nuevas rentas de bienes raíces en toda la ciudad corresponden a inmigrantes jubilados?

_____ %

*En las dos siguientes preguntas hablamos todo la propiedad inmobiliaria:*

6. Aproximadamente, ¿qué porcentaje de la propiedad inmobiliaria en el centro histórico de la ciudad es actualmente propiedad de inmigrantes de América del Norte o Europa?

_____ %

7. Aproximadamente, ¿qué porcentaje de la propiedad inmobiliaria en toda la ciudad pertenece actualmente a inmigrantes de América del Norte o Europa?

_____ %

8. ¿En general cómo difieren las preferencias de vivienda entre los inmigrantes jubilados y los residentes locales? En otras palabras, ¿qué cosas quieren los inmigrantes jubilados en una casa o apartamento que son diferentes a los residentes locales, y viceversa?

## Preguntas y observen en tiendas de barrio, pulperías / ventas, o recauderías

Para cada uno de los siguientes, díganos cuál es su precio actual y, según su mejor conocimiento, cuánto difiere de hace 5 años:

|  | Actualmente | Hace cinco años |
|---|---|---|
| a. Un litro de leche | _____ | _____ |
| b. Una docena de huevos | _____ | _____ |
| c. Un kilo / una libra de arroz | _____ | _____ |
| d. Un kilo / una libra de azúcar | _____ | _____ |
| e. Un kilo / una libra de cebollas | _____ | _____ |

Lista de Frutas y Verduras Frescas (FVF). Ponga una marca en el nombre de cada fruta que está disponible en la tienda

Notas:

|  | Disponible en la Tienda? | |
|---|---|---|
| **Frutas y Verduras** | **No** | **Sí** |
| a.  Plátano o banano (banana) | 0 | 1 |
| b.  Orito (plátano manzano)  (small or finger banana) | 0 | 1 |
| c.  Naranja (orange) | 0 | 1 |
| d.  Limón (lemon) | 0 | 1 |
| e.  Lima (lime) | 0 | 1 |
| f.  Manzana (apple) | 0 | 1 |
| g.  Melón (cantalope) | 0 | 1 |
| h.  Aguacate (avocado) | 0 | 1 |
| i.  Mango (mango) | 0 | 1 |
| j.  Cebolla (onion) | 0 | 1 |
| k.  Papa (potato) | 0 | 1 |
| l.  Tomate / jitomate (tomato) | 0 | 1 |
| m. Espárragos (asparagus) | 0 | 1 |
| n.  Fresas / frutillas (strawberries) | 0 | 1 |

# Appendix C: Interview Form for Local Residents – English Version

*My first questions are about how you feel that immigration of retired North Americans can and in the future could affect [Name of City], including the areas around the city.*

1. Would you say that, in your opinion, having immigrant retirees move to [NAME OF CITY] is good for the community, and if so, in what ways?

> Probe: Are there, in your opinion, drawbacks to having immigrant retirees living in [NAME OF CITY], and if so, what are some of those drawbacks?

2. What, in your opinion, have been the positive and negative effects that you are aware of, both positive and negative on quality of life of local citizens?

> Areas to probe: jobs, housing market and prices, transportation, the character of neighborhoods, the attitudes and activities of young people, and traditional attitudes.

3. Next, let's talk about the environment and issues of sustainable development. I would like to know what environmental issues are important in terms of maintaining environmental quality if the community continues to grow.

> Areas to probe: (a) water, (b) sewage, (c) transportation, and (d) public recreation areas such as parks?

4. Next, I would like to talk about issues of public safety and security. How has the movement of foreign retirees into [NAME OF CITY] had affected the security and safety of everyone living here?

> Probe: How has the city had to modify services because of the presence of foreigners?

5. One final question in this section. It's about differences within the native population. Are there ways that the impact of retiree migration may affect some but not others in your community?

> Probe: For example, perhaps negative impacts on the environment may be felt by all but economic benefits only by some. Please comment.

**Part 2**

*Now we are moving to the second part of the interview.*

1. How would you describe the relationship in general between local citizens and immigrant retirees? By this I mean the kinds of contact between immigrant retirees and the local community, and the type of relationships that develop?

2. What conflicts, if any, have you experienced between local customs and the activities of retired immigrants? Share an example if you have one and your thoughts about what communities can do to prevent or minimized such conflicts.

3. Some people think of immigrant retirees as a type of tourist. When you think about benefits and drawbacks to the community, how in your opinion, do immigrant retirees compare with more traditional tourists such as students or older persons who come for a short time and then go home?

4. What are your impressions on the integration of the retired immigrants here?

    a. Probe: Are there different subgroups or types of retired immigrants?
    b. Probe: If most retired immigrants were to live separately, socialize among their own kind, and learn little or no Spanish, would you consider this a problem? Why or why not?
    c. Probe: In your opinion, what kinds of actions of activities can best bring the two communities together?

5. Are there other things you would like to say about the presence of immigrant retirees in [NAME OF CITY]?

*If interviewee is a real estate agent, a government worker, or a convenience store / pulpería / venta / recaudería / abarrote owner, ask now the relevant questions from the last 2 pages of this form.*

**Part 3**

*We are now moving to the third part of this interview.*

Some Latin American countries or cities have laws that were designed to specifically attract immigrant retirees. Examples include relief from certain taxes and discounts on certain services.

1. What benefits are you aware of that [NAME OF COUNTRY] and [NAME OF CITY] currently offer to immigrant retirees, and how do you feel about providing those benefits?

2. One of the benefits that immigrant retirees often want is access to local health care. What is your opinion about whether immigrant retirees should be able to enroll in the government health services of [NAME OF COUNTRY] and in what circumstances should that occur.

3. If your community wanted to attract more immigrant retirees here, what additional things could the government do that in your opinion would be most useful?

## Part 4

*[Present laminated card with agreement response options]*

*Next I will say some statements about immigrant retirees. Please answer each by saying whether you completely agree (4), agree (3), neither agree nor disagree (2), disagree (1), or completely disagree (0). Remember – we are asking your own opinion as an individual and not as a representative of your work. Also, your responses will be completely confidential and will be combined with others to understand better how the community feels about these issues.*

| | 0 = Completeamente en desacuerdo<br>1 = en desacuerdo<br>2 = ni de acuerdo ni en desacuerdo<br>3 = de acuerdo<br>4 = completamente de acuerdo |
|---|---|
| 1. Having immigrant retirees move to [NAME OF CITY] is good because it creates jobs. | 0  1  2  3  4 |
| 2. Factories, farms, and other local businesses not linked to tourism have suffered because of the immigration of retirees. | 0  1  2  3  4 |
| 3. Real estate prices have increased because of the demand of retired immigrants. | 0  1  2  3  4 |
| 4. Real estate agents prefer to sell or rent to immigrant retirees. | 0  1  2  3  4 |
| 5. Foreigners are causing local residents to move from the center of the city to more outside areas of the city. | 0  1  2  3  4 |
| 6. Local residents who want to be successful in business need to learn English. | 0  1  2  3  4 |
| 7. Having Donald Trump as US president has not been good for local attitudes toward the U.S. and American retirees. | 0  1  2  3  4 |
| 8. Having immigrant retirees move here has led to more need for police and security services. | 0  1  2  3  4 |
| 9. Having a lot of retired foreigners in town is bad for the quality of life of the natives. | 0  1  2  3  4 |
| 10. Immigrant retirees are better for the community than tourists. | 0  1  2  3  4 |
| 11. The presence of so many North Americans has changed the culture of our city. | 0  1  2  3  4 |
| 12. Parts of our city are an American Colony. | 0  1  2  3  4 |
| 13. People who move here from other countries should learn Spanish | 0  1  2  3  4 |
| 14. Retired immigrants increase the cost of living for locals. | 0  1  2  3  4 |
| 15. Immigrant retirees often do not respect the local culture. | 0  1  2  3  4 |
| 16. North Americans do not understand the way we do things here. | 0  1  2  3  4 |
| 17. The presence of retired immigrants makes me feel like a second-class citizen in my own city | 0  1  2  3  4 |

| | |
|---|---|
| 18. The presence retired immigrants makes the rich richer but does not help the poor. | 0  1  2  3  4 |
| 19. The retired immigrants do charity and volunteer work that is valued by my community. | 0  1  2  3  4 |
| 20. Immigrant retirees act privileged when they are in public spaces. | 0  1  2  3  4 |
| 21. The immigrant retirees are a friendly and open in relation to natives of our city. | 0  1  2  3  4 |
| 22. The immigrant retirees are a bad influence on young people in our city. | 0  1  2  3  4 |
| 23. I would like to see more immigrant retirees move here to live. | 0  1  2  3  4 |

*We are almost done. I have two questions about you and then a final opportunity for you to say what is on your mind.*

1. How long have you lived in [NAME OF CITY]?                                    _____ years

2. Where in the city do you currently live? [Show a map of the city and ask the person to put a dot on that map]

3. What is your position?                    _____

4. How old are you in years?                                    _____ years

5. Is there anything else you would like to say from your perspective about anything we have talked about?

## Questions for Government Officials

1. What percent of the city's revenue comes from property taxes?

_____%

2. Is there a local strategic plan for addressing immigration?  Please describe it.

Probe:  If a copy is available, can you share it with us?

3. Have new programs been developed or changes in the city budget been instituted to better serve immigrant retirees?

4. Some people suggest that [NAME OF CITY] should charge more taxes that affect retired immigrants, such as real estate taxes, pay for better services to help the whole community, especially people in poorer areas.  Do you support this idea, and if so, what would you consider the spending priorities?

Probe:  What are the areas of greatest community need in terms of government services?

5. The lack of work opportunities for men is often a problem in communities that rely on tourism, including communities with large retiree populations.  Is this a problem in [NAME OF CITY] and, if so, what might be solutions that would help improve the situation?

6. How could the government of [NAME OF CITY] benefit more from immigration of retirees from North America?

## Questions for Real Estate Agents

1. Here is a map of [NAME OF CITY] divided into zones. For each zone, I would like for you to rate, using your knowledge of real estate, the amount of retired immigrants living there on a scale of 0 – 3, where  0 = none; 1 = few; 2 = some; 3 = many.

*The next two questions concern purchases of homes in the city.*

2. Approximately what percent of new real estate purchases in the historic city center are by retired immigrants?

_____%

3. Approximately what percent of real estate purchases in [NAME OF CITY] overall are by retired immigrants?

_____%

*The next two questions concern rental homes in the city.*

4. Approximately what percent of real estate rentals in the historic city center are by retired immigrants?

_____%

5. Approximately what percent of real estate rentals in [NAME OF CITY] overall are by retired immigrants?

_____%

*The next two questions concern all property homes in the city.*

6. Approximately what percent of the real estate in the historic city center is currently owned by immigrants from North America or Europe?

_____%

7. Approximately what percent of the real estate in [NAME OF CITY] overall is currently owned by immigrants from North America or Europe?

_____%

8. How do housing preferences in general differ between immigrant retirees and local residents? In other words, what things to immigrant retirees want in a house or apartment that are different than local residents, and the other way around?

## Questions and Observations for Convenience Store / Pulpería / Venta / Recaudería / Tienda de Barrio Owners

For each of the following, tell us what your current price is and, to the best of your knowledge, how different it was 5 years ago:

|  | Now | Five Years Ago |
|---|---|---|
| a. A liter of milk | _____ | _____ |
| b. A dozen eggs | _____ | _____ |
| c. A kilo / a liter of rice | _____ | _____ |
| d. A kilo / a liter of sugar | _____ | _____ |
| e. A kilo / a liter of onions | _____ | _____ |

Checklist of Fresh Fruit and Vegetables (FFVs). Put a check by the name of each fruit that is available in the store today

Notas:

| | Disponible en la Tienda? | |
|---|---|---|
| **Frutas y Verduras** | **No** | **Sí** |
| a.  Plátano o banano (banana) | 0 | 1 |
| b.  Orito (plátano manzano)  (small or finger banana) | 0 | 1 |
| c.  Naranja (orange) | 0 | 1 |
| d.  Limón (lemon) | 0 | 1 |
| e.  Lima (lime) | 0 | 1 |
| f.  Manzana (apple) | 0 | 1 |
| g.  Melón (cantalope) | 0 | 1 |
| h.  Aguacate (avocado) | 0 | 1 |
| i.  Mango (mango) | 0 | 1 |
| j.  Cebolla (onion) | 0 | 1 |
| k.  Papa (potato) | 0 | 1 |
| l.  Tomate / jitomate (tomato) | 0 | 1 |
| m.  Espárragos (asparagus) | 0 | 1 |
| n.  Fresas / frutillas (strawberries) | 0 | 1 |

# Appendix D: Coding System Used for Analyzing Interview Data

Codes for Interviews and Field Notes

| | |
|---|---|
| **1. Retirees**<br>    A. Characterization<br>    B. Motivation<br>    C. Activities<br>    D. First Impression<br>    E. Philanthropy<br>    F. Health Status<br>    G. Tourist vs retiree<br>    H. Positive<br>    I. Negative | **9. Government / Service Providers**<br>    A. Benefits offered<br>    B. Current State<br>    C. Lack of knowledge<br>    D. Attitudes toward retiree use<br>    E. Non governmental initiatives |
| **2. Local/Native Residents**<br>    A. Characterization<br>    B. Activities<br>    C. Positive<br>    D. Negative | **10. Health Care**<br>    A. Current State<br>    B. Attitude toward retiree use<br>    C. Quality<br>    D. Cost |
| **3. Culture**<br>    A. Similarities<br>    B. Differences | **11. Recommendations**<br>    A. Recs for Integration<br>    B. Deter Expat Migration |
| **4. Environment (including open space, water, food, transportation)**<br>    A. Current State<br>    B. Retiree relation to<br>    C. Anticipated Change | **12. Migration**<br>    A. Returning<br>    B. Latin American<br>    C. Internal Migrants<br>    D. Migration Attitudes |
| **5. Safety / Security**<br>    A. Current State<br>    B. Retiree relation to | **13. Drugs & Organized Crime** |
| **6. Real Estate**<br>    A. Current State<br>    B. Retiree relation to<br>    C. Expat preferences<br>    D. Displacement | **14. Communication / Language**<br>    A. Communication Barrier<br><br>**15. Pride** |
| **7. Economy / Jobs**<br>    A. Current State<br>    B. Retiree relation to | **16. Identity**<br><br>**17. Prejudice**<br>    A. Participant Prejudice |
| **8. Social Dynamics**<br>    A. Harmony<br>    B. Conflict<br>    C. Segregation<br>    D. Integration<br>    E. Paid relationship<br>    F. Relationships<br>    G. Power/Privilege | **18. Quotes**<br><br>**19. Stories**<br><br>**20. Unsure Code** |

| | Code | Description and Decision Rule(s) |
|---|---|---|
| **1. Retirees** | **1. Retirees** | No coding under 1. Retirees, use codes below. |
| | **Characterization** | Descriptions of expats: ex. always on-time, friendly, unfriendly, loud, etc. How are local residents and others (including expats!) characterizing retirees in their behavior, personalities, mannerisms. These statements can be neutral or charged, as long as it is a description of expats as a group. |
| | **Motivation** | What are the motivations for retirees are to move to SMA/Cuenca? |
| | **Activities** | Whenever a respondent makes mention of an activity that they attribute fairly specifically to retirees. Do not include philanthropic/volunteering activities in this code. |
| | **First impression** | What are the first things that local residents have come to mind when asked about retirees? Should be consistently coded on the first/second question of interview, even if this is "nothing", etc. |
| | **Philanthropy** | Any mentions of expats/retirees engaging in philanthropic efforts – volunteering, attending charity events, working with NGOs and other non profits. |
| | **Health Status** | References to retiree's health – ability vs. disability, mobility, diseases, etc. Not references to use of health care (which is under benefits). |
| | **Tourist vs retiree** | Descriptions of differences in behaviors, activities or impacts that retirees have versus traditional tourists. |
| | **Positive** | These will often be double coded! Use anytime an interviewee makes a positive/negative reference SPECIFICALLY and EXPLICITLY about retirees. This will be able to apply to any other subject (i.e. environment, culture, etc.) as long as they are making a specific statement about whether it is a positive or negative thing about retirees. |
| | **Negative** | |
| **2. Local Residents** | **2. Local Residents** | No coding under 2. Local/Native Residents, use codes below. |
| | **Characterization** | Descriptions of local/native residents as time-flexible, friendly, unfriendly, loud, etc. How are interviewees characterizing local residents in their behavior, personalities, mannerisms, etc.? This can include descriptions of cultural conventions as well. Can come both from retirees/expats, but also from local residents speaking about local residents as a group. |
| | **Activities** | Whenever an interviewee makes mention of an activity that they attribute fairly specifically to Cuencanos/San Miguelers. This can include descriptions of activities typically attributed to local. |
| | **Positive** | These will often be double coded! Use anytime a resident makes a positive/negative reference SPECIFICALLY about local residents. This will be able to apply to any other subject (i.e. environment, culture, etc.) as long as they are making a specific statement on whether it is a positive or negative thing about local residents. |
| | **Negative** | |
| **3. Culture** | **3. Culture** | No coding under 3. Culture, use codes below. |
| | **Similarities** | Cultural similarities between retirees & local residents. |
| | **Differences** | Mentions of cultural differences between retirees & local residents. Any mention of cultural differences between retirees (whether U.S., Canadian, or other) and Latin American culture. Can either be positive or negative – use other positive/negative codes if applicable. |
| **4. Environment** | **4. Environment** | No coding under 4. Environment, use codes below. |
| | **Current State** | Any statements about the current status of water access, open public space access, sewer/sanitation systems, food, transportation systems, other environmental and urban environmental factors. This includes past changes that have brought about the current status of environmental systems in Cuenca/San Miguel de Allende. These statements can be general and don't need to relate to retirees or migration. |
| | **Retiree Relation** | Statements specifically concerning retirees and their relation to the environment. |
| | **Anticipated Change** | Any change that is anticipated, or respondents *think* will happen in the future. Primarily in response to question that prompts for respondents to think about future state of the environment. |
| **5. Safety/Security** | **5. Safety / Security** | No coding under 5. Safety / Security, use codes below. |
| | **Current State** | Any statements about the current status of public safety and security services in the city. This includes past changes that have brought about the current status. These statements can be general, and don't need to relate to retirees or migration. |
| | **Retiree relation** | Statements specifically concerning retirees and their relation to the public safety/security of the city. |
| **6. Real Estate** | **6. Real Estate** | No coding under 6. Real Estate, use codes below. |
| | **Current State** | Any statements about the current status of real estate conditions in the city. This includes changes that have brought about the current status. |
| | **Retiree relation** | Statements specifically concerning retirees and their impact/relation to real estate. |
| | **Expat preferences** | Any note on expat preferences versus local residents' preferences (whether explicit or implicit). |
| | **Displacement** | Gentrification/displacement of local residents from areas. Any references to local residents selling or moving to make space for expats OR references to locals actively not selling (phenomenon of gentrification & resistance). |

| | | |
|---|---|---|
| **7. Economy/Jobs** | **7. Economy / Jobs** | No coding under 7. Economy / Jobs, use codes below. |
| | **Current State** | Any statements about the current status of the economy. This includes changes that have brought about the current status. |
| | **Retiree relation** | Statements specifically concerning retirees and their relation to the economy. General economic statements about the city that do not fall under real estate or government. |
| **8. Social Dynamics** | **8. Social Dynamics** | No coding under 8. Social Dynamics, use codes below. |
| | **Harmony** | Examples of harmonious living between expatriates and local residents – must be within the group setting (see below for individual relationships). This can be either integrated harmony or segregated harmony. |
| | **Conflict** | Examples of conflict between residents & expatriates. This includes both one-one-one conflicts (specific relationships/examples) or more general, group conflicts. Can be talking about conflict both as they are integrating, or conflict in the fact that there is not much integration - will be double coded as such. |
| | **Segregation** | Any example of local residents saying that expats live separately, or don't interact, or they do completely different things. Can either be a positive or negative connotation or interaction (see above for double coding with Harmony/ Conflict). |
| | **Integration** | Examples of local residents talking about expats integrating into the culture, communities, etc. Can either be a positive or negative connotation or interaction (see above for double coding with Harmony/Conflict). |
| | **Paid** | Examples of expat and local resident relationships that are paid & include monetary transactions. Any paid (whether purely business or not) transaction will be included. This can be doctor appointments, but also close relationships with paid help in the house or other facilitator relationships. |
| | **Relationships** | Warm, personal, positive one-on-one relationships. MUST be intercultural, between expats & local residents. Runs the gamut of close friends, and personal acquaintances, as long as is a mention of a particular relationship rather than just general group interaction between expats/local residents. |
| | **Power/Privilege** | Mentions of power dynamics between foreign community & local resident community. This includes: purchasing power differentials, knowledge power, privilege, social capital power, agency and freedom in environment, influence, etc. This code applies to both perceived & real. |
| **9. Government** | **9. Government** | No coding under 9. Government, use codes below |
| | **Benefits** | Any statements about offered benefits (whether true or not). This does not apply to statements about retiree's use of benefits. |
| | **Current State** | Any statement about the current state of the government (except for as it relates to benefits). Information about taxes, existing government policies, politics, etc. |
| | **Lack of Knowledge** | A statement that explicitly states "I'm not sure what is there, or what the status of XYZ is". |
| | **Attitudes** | Attitude statements about the state of benefits as they relate to expats. |
| | **Non-Governmental Initiatives** | Existing or suggested initiatives put into place by service providers, non profits, and any other non-governmental organization that are aimed at managing incoming retirees (either attracting retirees, or managing in some way). |
| **10. Health Care** | **10. Health Care** | No coding under 10. Health Care, use codes below |
| | **Current State** | Any statements about the offered health care, whether private or public. |
| | **Attitude toward use** | Judgement statements on retiree's use of health care benefits. |
| | **Quality** | Any reference to the quality of health care (whether bad or good). |
| | **Cost** | Any reference to the cost of health care (whether bad or good). |
| **11. Recommendations** | **11. Recommendations** | *PARENT CODE THAT YOU CAN CODE IN* (see specifics below) |
| | **Recommendations** | Any informal recommendations that are made by interviewees. |
| | **Recommendations for Integration** | Any plans or suggestions for social integration between groups – both formal & informal. Ex. Starting a pot luck at a church, or having required language classes to encourage communication. |
| | **Deter Expat Migration** | Interviewees references the desire to reduce/stop retirees from moving into the city. |
| **12. Migration** | **12. Migration** | No coding under 12. Migration, use codes below |
| | **Returning** | References to returning migrants as a phenomenon. This is particularly prevalent in Cuenca. |
| | **Latin American** | References to other migrants from nearby countries. |
| | **Internal Migrants** | References to migrants from within country (Ecuador to Cuenca or Mexico to SMA). |
| | **Migration Attitudes** | Attitudes towards migration – charged, and have a judgment. Examples: attitudes towards migrants from other LA countries and attitudes about immigration policy in the States. |

| | | |
|---|---|---|
| **13. Drugs & Crime** | **13. Drugs & Organized Crime** | *PARENT CODE THAT YOU CAN CODE IN* |
| | **Drugs & Organized Crime** | Use of drugs / alcohol; as well as distribution of drugs / alcohol. Covers illegal activity – although some legal as well (when talking about legal substance use, ex. Alcohol consumption). Other types of crime activity as well – guns, trafficking, etc. |
| **14. Comm/Language** | **14. Communication/ Language** | *PARENT CODE THAT YOU CAN CODE IN* |
| | **Communication/ Language** | Encompasses references of both non-verbal & verbal communication/language. |
| | **Communication Barrier** | Specific barriers due to breakdowns in communication and language – whether due to retiree's lack of Spanish, or their lack of knowledge of common nonverbal communication expectations in Cuenca/San Miguel. |

| | |
|---|---|
| **THE FOLLOWING ARE IMPLICIT CODES – CODES THAT USE CODER'S JUDGEMENT TO A GREATER EXTENT** | |
| **15. Pride** | *PARENT CODE THAT YOU CAN CODE IN* |
| **Pride** | References to a group identity, etc. Both implicit and explicit examples of this – when the coder reads/feels that the interviewee is prideful of themselves, their community, their business, etc., but also when it is specifically mentioned. |
| **16. Identity** | *PARENT CODE THAT YOU CAN CODE IN* |
| **Identity** | References to interviewees' identity – "I identify as_____"or "XYZ is part of me". This is specific to group identities – city, state, expat group, etc. rather than identifying as a mother, father, etc. |
| **17. Prejudice** | *PARENT CODE THAT YOU CAN CODE IN* |
| **Prejudice** | This code is for when interviewees brings up examples of prejudice occurring, whether it is outright statement or a bit more subtle. This includes examples of prejudice *against* the interviewee, and examples of general prejudice. Prejudice includes examples of negative stereotyping. Can be prejudice against local residents or retirees. |
| **Participant Prejudice** | This is prejudice that the **CODER** is interpreting – the **CODER** feels that the participant/interviewee has made a statement that indicates prejudice on the interviewee's behalf. Reserved for strong statements (ex. [this group] are dishonest, bad people). The interviewee themselves are making negative stereotypes or strong statement against either local residents or retirees. |
| **18. Quote** | *PARENT CODE THAT YOU CAN CODE IN* |
| **Quote** | ANY good quotes (should be short, a few lines at most.) These can be on any subject, negative or positive. |
| **19. Stories** | *PARENT CODE THAT YOU CAN CODE IN* |
| **Stories** | ANY good stories. These are longer, and not necessarily a good 'quote' in it, but the gist of the entire story as a unit is worth keeping a code on. |
| **20. Unsure** | *PARENT CODE THAT YOU CAN CODE IN* |
| **Unsure** | Unsure – seems important, but not sure where it would fit in within our existing code book. PLEASE add an annotation to these, to say a little while you think it is important/make a note about it. |

# Appendix E: Online Survey of Immigrant Retirees

| Part 1 |
|:---:|

*Eligibility*

1. Are you a native of the United States, Canada, or a country in Europe who has moved to Ecuador after age 50?

    ☐ Yes
    ☐ No
Please enter your native country: _____

2. Are you 55 years of age or older?

    ☐ Yes
    ☐ No

3. Did you spend at least 4 months as a resident of [CITY] in the past 12 months?

    ☐ Yes
    ☐ No

**If Non-Eligible:**

"We are sorry! You are not eligible for this survey. Thank you for your interest. If you know people who qualify, we would appreciate your letting them know about the survey."

*General Information*

4. Please indicate your sex:

    ☐ Male
    ☐ Female
    ☐ Prefer not to answer

5. How old are you in years? _____

6. What is your current visa status in [CITY]? Please check the most appropriate response.

    ☐ Permanent Resident
    ☐ Temporary Resident
    ☐ Tourist Visa/Visitor Permit
    ☐ Citizen
    ☐ Other (please describe): _____

7. How long have you had a temporary or permanent residence in [CITY]?

      Years: _____

      Months: _____

8. How many months in the past year have you spent in [COUNTRY]?

  0    1    2    3    4    5    6    7    8    9    10    11    12

9. How many times in the past year did you travel back to your country of origin?

    ☐ Not at all
    ☐ Once
    ☐ Twice
    ☐ Three times or more

10. Map

    On the below map [zoomed in to either CUENCA or SAN MIGUEL DE ALLENDE], please drag and drop the marker to where you live as best as you can. This information is kept confidential. You are able to zoom in and out by either 1) double clicking or 2) using the +/- sign at the bottom right corner.

*Your Experiences Moving to and Living in [CITY]*

11. What was your primary motivation for moving to [CITY]?

    ☐ Lifestyle
    ☐ Climate
    ☐ Get away from home country
    ☐ Affordable retirement
    ☐ Affordable health care
    ☐ Work
    ☐ Other (please specify): _____

12. What factor was most important in your decision to move to [CITY]?

    ☐ The advice of friends already living here
    ☐ Research from books or magazines
    ☐ Research from websites/blogs
    ☐ Previous travel to {NAME OF COUNTRY}
    ☐ Other (please specify): _____

13. My closest friend in [CITY] is a native of:

☐ United States
☐ Canada
☐ [CITY]
☐ Other (please specify):_____

*Safety Net*

The next few questions are about the people who are available to help you if you have a problem. For each question, you have the same response options.

14. If you were sick with a fever and needed someone to take you to a doctor or get medicine, who would you most likely call first?

☐ Family member or friend who lives with you
☐ Friend who lives in [CITY] and is from your home country
☐ Friend who is a native of [CITY]
☐ Manager of the apartment or neighborhood where you live
☐ Local resident who works for you
☐ Friend or family who live in your country of origin
☐ Other (please describe): _____

15. If you needed a place to stay temporarily because your home had become flooded by a leaking water pipe, who would you most likely call first?

☐ Family member or friend who lives with you
☐ Friend who lives in [CITY] and is from your home country
☐ Friend who is a native of [CITY]
☐ Manager of the apartment or neighborhood where you live
☐ Local resident who works for you
☐ Friend or family who live in your country of origin
☐ Other (please describe): _____

16. If you lost your wallet and needed to borrow money on a weekend, who would you most likely call first?

☐ Family member or friend who lives with you
☐ Friend who lives in [CITY] and is from your home country
☐ Friend who is a native of [CITY]
☐ Manager of the apartment or neighborhood where you live
☐ Local resident who works for you
☐ Friend or family who live in your country of origin
☐ Other (please describe):

17. How would you rank your Spanish skills?

- ☐ Fluent
- ☐ Very good
- ☐ Simple conversations
- ☐ Limited
- ☐ None

18. Which best describes your attitude toward improving your Spanish skills?

- ☐ I am already fluent in reading and speaking Spanish
- ☐ I am interested in improving my language skills to the point of having extensive conversations easily in Spanish
- ☐ I am interested in improving my language skills to the point of conducting business transactions easily in Spanish
- ☐ I would rather conduct my life in English as much as possible rather than learn another language

The next questions are about the neighbors who live in the two units closest to your house or apartment.

19. Where is closest neighbor #1 from (born or spent much of their life):

- ☐ I don't know
- ☐ [CITY]
- ☐ USA
- ☐ Canada
- ☐ Other: _____

20. Where is closest neighbor #2 from (born or spent much of their life):

- ☐ I don't know
- ☐ [CITY]
- ☐ USA
- ☐ Canada
- ☐ Other: _____

*Health Care/Community Involvement*

21. What percentage of your health care did you get in Ecuador in the last year?

0      10      20      30      40      50      60      70      80      90      100

22. Which of these describes something you have done in the past year in [CITY]?

☐ Pay real estate taxes
☐ Develop or invest in a business within the city
☐ Help a local family financially (ex. help with school fees, or paying for needed medicine). Please explain:
☐ Work for a local charity organization. Please explain: _____
☐ Other. Please explain: _____

*Benefits Provided to You by the Government of [CITY]*

23. Here is a list of things some Latin American countries or cities provide for foreign retirees. For each, state if it is available in [CITY] and, if it is, your opinion about how useful and important it is in attracting retirees or making them happy to live here.

| | Available in CITY? | | | Importance to You if Available | | |
|---|---|---|---|---|---|---|
| | No | Yes | Don't Know | Little or No Benefit | Some Benefit | Much Benefit |
| a. Exemption from taxes on money brought into the country for purchasing, building, or restoring real estate | ☐ | ☐ | ☐ | ☐ | ☐ | ☐ |
| b. Exemption from taxes from bringing a car into the country | ☐ | ☐ | ☐ | ☐ | ☐ | ☐ |
| c. Exemption from paying taxes on social security or other pension income from their home country | ☐ | ☐ | ☐ | ☐ | ☐ | ☐ |
| d. Exemptions or discounts for certain municipal taxes, such as selling property | ☐ | ☐ | ☐ | ☐ | ☐ | ☐ |
| e. Access to the government health care system after a waiting period. | ☐ | ☐ | ☐ | ☐ | ☐ | ☐ |
| f. Discounts on utilities, such as telephone, internet, television, and drinking water. | ☐ | ☐ | ☐ | ☐ | ☐ | ☐ |
| g. Discounts on public transportation. | ☐ | ☐ | ☐ | ☐ | ☐ | ☐ |
| h. Discounts on groceries. | ☐ | ☐ | ☐ | ☐ | ☐ | ☐ |
| i. Discounts on entertainment. | ☐ | ☐ | ☐ | ☐ | ☐ | ☐ |
| j. Maintenance of parks and other recreational opportunities. | ☐ | ☐ | ☐ | ☐ | ☐ | ☐ |
| k. Provision of free classes in Spanish or opportunities/ "exchanges" where people who are learning Spanish and English can talk to each other. | ☐ | ☐ | ☐ | ☐ | ☐ | ☐ |

24. For each of the following statements, please choose the response that most represents your opinion or feeling:

| | Strongly Agree | Agree | Somewhat Agree | Neither Agree nor Disagree | Somewhat Disagree | Disagree | Strongly Disagree |
|---|---|---|---|---|---|---|---|
| a. I moved to [CITY] because I was dissatisfied with my home country. | ☐ | ☐ | ☐ | ☐ | ☐ | ☐ | ☐ |
| b. I moved to [CITY] because I was unhappy with the politics back home. | ☐ | ☐ | ☐ | ☐ | ☐ | ☐ | ☐ |
| c. My decision to move to [CITY] was mainly economic. | ☐ | ☐ | ☐ | ☐ | ☐ | ☐ | ☐ |
| d. I live here because property taxes are lower. | ☐ | ☐ | ☐ | ☐ | ☐ | ☐ | ☐ |
| e. I consider [CITY] to be my home. | ☐ | ☐ | ☐ | ☐ | ☐ | ☐ | ☐ |
| f. I feel at home in [CITY]. | ☐ | ☐ | ☐ | ☐ | ☐ | ☐ | ☐ |
| g. [CITY] feels more like a place where I live rather than a part of who I am. | ☐ | ☐ | ☐ | ☐ | ☐ | ☐ | ☐ |
| h. I am satisfied with my life here in [CITY]. | ☐ | ☐ | ☐ | ☐ | ☐ | ☐ | ☐ |
| i. I worry that too many Americans are ruining the [CITY]. | ☐ | ☐ | ☐ | ☐ | ☐ | ☐ | ☐ |
| j. I don't have a political voice living here. | ☐ | ☐ | ☐ | ☐ | ☐ | ☐ | ☐ |
| k. I live here because the health care is affordable. | ☐ | ☐ | ☐ | ☐ | ☐ | ☐ | ☐ |
| l. I live here because I can get better, cheaper care if I have a disability or develop Alzheimer's disease. | ☐ | ☐ | ☐ | ☐ | ☐ | ☐ | ☐ |
| m. I plan to continue to live in [CITY] even if I become disabled. | ☐ | ☐ | ☐ | ☐ | ☐ | ☐ | ☐ |
| n. I am interested in developing friendship with people who are part of the native community. | ☐ | ☐ | ☐ | ☐ | ☐ | ☐ | ☐ |
| o. I am interested in developing friendships with people who are part of the native community. | ☐ | ☐ | ☐ | ☐ | ☐ | ☐ | ☐ |
| p. I am more comfortable spending time with people from my own culture. | ☐ | ☐ | ☐ | ☐ | ☐ | ☐ | ☐ |
| q. I love to study and learn about local language and culture. | ☐ | ☐ | ☐ | ☐ | ☐ | ☐ | ☐ |
| r. I attend events that are put on by the local community and conducted in Spanish. | ☐ | ☐ | ☐ | ☐ | ☐ | ☐ | ☐ |
| s. I feel like I belong in [CITY]. | ☐ | ☐ | ☐ | ☐ | ☐ | ☐ | ☐ |
| t. I feel like the residents in [CITY] accept me. | ☐ | ☐ | ☐ | ☐ | ☐ | ☐ | ☐ |
| u. I have meaningful friendships with people who are part of the native community. | ☐ | ☐ | ☐ | ☐ | ☐ | ☐ | ☐ |

*We are almost finished.*

25. Please tell us about an experience you have had or an observation you've made that illustrates something that you consider important for others to know about living in [CITY].

Click or tap here to enter text.

26. In your opinion, what are the best things that a community can do to attract foreign retirees. Please list up to three things.

1.

2.

3.

27. In your opinion, what are three potential negative impacts of retirement migration on [CITY]?

1.

2.

3.

*About You*

28. What is your highest level of education?

☐ I did not complete high school
☐ High school graduate
☐ 2-year college
☐ 4-year college
☐ Post college

29. Please indicate your current employment status:

☐ Fully retired
☐ Part-time employment
☐ Full-time employment

30. What is your monthly income (in USD)?

☐ Less than $1,000 per month
☐ $1,000 - $3,000 per month
☐ Greater than $3,000 per month

31. With whom do you live? (you can answer more than one):

- ☐ Alone
- ☐ Spouse or Partner
- ☐ Other family member
- ☐ Friend
- ☐ Roommate (a person you house with, but are not otherwise connected to)
- ☐ Other: _____

32. Please choose the answer that best describes your current living situation in [CITY].

- ☐ Hotel
- ☐ Rental apartment or town house
- ☐ Apartment or town house I own
- ☐ House I rent
- ☐ House I own
- ☐ Other (please specify): _____

33. Do you live in a community that has controlled entry by a gate and/or guard?

- ☐ Yes
- ☐ No

34. Which one of the following best describes what you do with your cash savings (bank accounts, certificates of deposit):

- ☐ I have some cash or no cash in a bank in [CITY], but keep most of my cash savings in my home country
- ☐ I have accounts in both countries, and with approximately equal amounts in each
- ☐ I have most or all of my cash savings in one or more banks in [CITY]
- ☐ Other - please describe: _____

Thank you for taking the time to complete this survey! Your response will be a great help in further understanding retirement to foreign countries, and how this affects retirees as well as the cities themselves. If you are interested in obtaining a copy of the finished report from this study, please check the box and enter your email below.

If you have any comments on the survey or the project, feel free to leave a comment below as well!

- ☐ Yes, I would like to receive a copy of the report. Add your email: _____

Any additional comments? Feel free to leave them below:

# Appendix F: Interview Form for Immigrant Retirees Using Long-Term Care

## Part 1

Did you complete our online survey?

If no → complete the entire interview                                          ☐
If yes → continue but do not complete the supplemental questions              ☐

Our research group is interested in the availability and quality of health and long-term care services in [COUNTRY] for persons like yourself. So, I am interested in your perspectives on the options both here and in your home country.

1. Tell me what you know about the availability and cost of personal care services <u>in the home or in a group setting such as assisted living</u>, for help with such things as bathing, dressing, getting to and from the bathroom, using the bathroom, and assistance with taking medications. What do you know about the options here?

2. Now tell me about <u>your personal experience</u> using these services.

   Probe:   Which services have you personally used?
            What have you heard about other people's experiences with these services?

3. If you were in your home country, how would the selection, availability, and cost of long-term care service differ from what they are here?

4. Are you currently receiving any of the following health services in [Cuenca]?

   Home visits by a doctor   $\square_0$

   Help with cooking, cleaning or shopping by someone you pay   $\square_1$

   Help from a nurse or nurses aid who comes to your home   $\square_2$

   A place you go regularly that provides meals and health care (day program for adults)   $\square_3$

   A place you live in that provides a bed, meals, and health care (assisted living or nursing home)   $\square_4$

5. To what extent, to your knowledge, are there any current or planned buildings in this community for provision of long-term care services to retired foreign immigrants? I'm talking about such as assisted living settings, nursing homes, or group homes that cater to retired persons and admit non-natives like yourself.

6. Are there certain health conditions under which you would return to your home country?
   Probe: Please describe.

7.  Next I want to ask a few questions about what you would do if your health got worse.  For each question, I would like to ask you to tell me whether you agree, somewhat agree, are unsure, somewhat disagree, or disagree.  [Show copy of the response options]

|  | Disagree (0) | Somewhat disagree (1) | Unsure (2) | Somewhat agree (3) | Agree (4) |
|---|---|---|---|---|---|
| If I were diagnosed with cancer I would remain in Ecuador for treatment | 0 | 1 | 2 | 3 | 4 |
| If I had a stroke with disability I would remain in Ecuador for treatment | 0 | 1 | 2 | 3 | 4 |
| If I had severe disability I would remain in Ecuador | 0 | 1 | 2 | 3 | 4 |
| If I required nursing home care I would remain in Ecuador | 0 | 1 | 2 | 3 | 4 |
| If I required home help with cooking , cleaning and bathing I would remain in Ecuador | 0 | 1 | 2 | 3 | 4 |

8.  Please tell me about your plans for what you would do if you became disabled or too ill to take care of yourself.

9.  Are there certain health conditions under which you would return to your home country?
    Probe:  Please describe

10. What health or personal care services are not available here that you wish were available.

**[Go to supplemental questions if the person has not completed the online form.
Otherwise say "That's all.  Thank you so much!]**

## Part 2

*Eligibility*

1. Are you a native of the United States, Canada, or a country in Europe who has moved to Ecuador after age 50?

☐ Yes
☐ No
Please enter your native country: _____

2. Are you 55 years of age or older?

☐ Yes
☐ No

3. Did you spend at least 4 months as a resident of [CITY] in the past 12 months?

☐ Yes
☐ No

**If Non-Eligible:**

"We are sorry! You are not eligible for this survey. Thank you for your interest. If you know people who qualify, we would appreciate your letting them know about the survey."

*General Information*

4. Please indicate your sex:

☐ Male
☐ Female
☐ Prefer not to answer

5. How old are you in years? _____

6. What is your current visa status in [CITY]? Please check the most appropriate response.

☐ Permanent Resident
☐ Temporary Resident
☐ Tourist Visa/Visitor Permit
☐ Citizen
☐ Other (please describe): _____

7. How long have you had a temporary or permanent residence in [CITY]?

    Years: _____

    Months: _____

8. How many months in the past year have you spent in [COUNTRY]?

  0    1    2    3    4    5    6    7    8    9    10    11    12

9. How many times in the past year did you travel back to your country of origin?

    ☐ Not at all
    ☐ Once
    ☐ Twice
    ☐ Three times or more

10. Map

    On the below map [zoomed in to either CUENCA or SAN MIGUEL DE ALLENDE], please drag and drop the marker to where you live as best as you can. This information is kept confidential. You are able to zoom in and out by either 1) double clicking or 2) using the +/- sign at the bottom right corner.

*Your Experiences Moving to and Living in [CITY]*

11. What was your primary motivation for moving to [CITY]?

    ☐ Lifestyle
    ☐ Climate
    ☐ Get away from home country
    ☐ Affordable retirement
    ☐ Affordable health care
    ☐ Work
    ☐ Other (please specify): _____

12. What factor was most important in your decision to move to [CITY]?

    ☐ The advice of friends already living here
    ☐ Research from books or magazines
    ☐ Research from websites/blogs
    ☐ Previous travel to {NAME OF COUNTRY}
    ☐ Other (please specify): _____

13. My closest friend in [CITY] is a native of:

☐ United States
☐ Canada
☐ [CITY]
☐ Other (please specify):_____

*Safety Net*

The next few questions are about the people who are available to help you if you have a problem. For each question, you have the same response options.

14. If you were sick with a fever and needed someone to take you to a doctor or get medicine, who would you most likely call first?

☐ Family member or friend who lives with you
☐ Friend who lives in [CITY] and is from your home country
☐ Friend who is a native of [CITY]
☐ Manager of the apartment or neighborhood where you live
☐ Local resident who works for you
☐ Friend or family who live in your country of origin
☐ Other (please describe): _____

15. If you needed a place to stay temporarily because your home had become flooded by a leaking water pipe, who would you most likely call first?

☐ Family member or friend who lives with you
☐ Friend who lives in [CITY] and is from your home country
☐ Friend who is a native of [CITY]
☐ Manager of the apartment or neighborhood where you live
☐ Local resident who works for you
☐ Friend or family who live in your country of origin
☐ Other (please describe): _____

16. If you lost your wallet and needed to borrow money on a weekend, who would you most likely call first?

☐ Family member or friend who lives with you
☐ Friend who lives in [CITY] and is from your home country
☐ Friend who is a native of [CITY]
☐ Manager of the apartment or neighborhood where you live
☐ Local resident who works for you
☐ Friend or family who live in your country of origin
☐ Other (please describe):

17. How would you rank your Spanish skills?

☐ Fluent
☐ Very good
☐ Simple conversations
☐ Limited
☐ None

18. Which best describes your attitude toward improving your Spanish skills?

☐ I am already fluent in reading and speaking Spanish
☐ I am interested in improving my language skills to the point of having extensive conversations easily in Spanish
☐ I am interested in improving my language skills to the point of conducting business transactions easily in Spanish
☐ I would rather conduct my life in English as much as possible rather than learn another language

The next questions are about the neighbors who live in the two units closest to your house or apartment.

19. Where is closest neighbor #1 from (born or spent much of their life):

☐ I don't know
☐ [CITY]
☐ USA
☐ Canada
☐ Other: _____

20. Where is closest neighbor #2 from (born or spent much of their life):

☐ I don't know
☐ [CITY]
☐ USA
☐ Canada
☐ Other: _____

*Health Care/Community Involvement*

21. What percentage of your health care did you get in Ecuador in the last year?

0    10    20    30    40    50    60    70    80    90    100

22. Which of these describes something you have done in the past year in [CITY]?

    ☐ Pay real estate taxes
    ☐ Develop or invest in a business within the city
    ☐ Help a local family financially (ex. help with school fees, or paying for needed medicine).
    Please explain:
    ☐ Work for a local charity organization. Please explain: _____
    ☐ Other. Please explain: _____

*Benefits Provided to You by the Government of [CITY]*

23. Here is a list of things some Latin American countries or cities provide for foreign retirees. For each, state if it is available in [CITY] and, if it is, your opinion about how useful and important it is in attracting retirees or making them happy to live here.

| | Available in CITY? | | | Importance to You if Available | | |
|---|---|---|---|---|---|---|
| | No | Yes | Don't Know | Little or No Benefit | Some Benefit | Much Benefit |
| a. Exemption from taxes on money brought into the country for purchasing, building, or restoring real estate | ☐ | ☐ | ☐ | ☐ | ☐ | ☐ |
| b. Exemption from taxes from bringing a car into the country | ☐ | ☐ | ☐ | ☐ | ☐ | ☐ |
| c. Exemption from paying taxes on social security or other pension income from their home country | ☐ | ☐ | ☐ | ☐ | ☐ | ☐ |
| d. Exemptions or discounts for certain municipal taxes, such as selling property | ☐ | ☐ | ☐ | ☐ | ☐ | ☐ |
| e. Access to the government health care system after a waiting period. | ☐ | ☐ | ☐ | ☐ | ☐ | ☐ |
| f. Discounts on utilities, such as telephone, internet, television, and drinking water. | ☐ | ☐ | ☐ | ☐ | ☐ | ☐ |
| g. Discounts on public transportation. | ☐ | ☐ | ☐ | ☐ | ☐ | ☐ |
| h. Discounts on groceries. | ☐ | ☐ | ☐ | ☐ | ☐ | ☐ |
| i. Discounts on entertainment. | ☐ | ☐ | ☐ | ☐ | ☐ | ☐ |
| j. Maintenance of parks and other recreational opportunities. | ☐ | ☐ | ☐ | ☐ | ☐ | ☐ |
| k. Provision of free classes in Spanish or opportunities/ "exchanges" where people who are learning Spanish and English can talk to each other. | ☐ | ☐ | ☐ | ☐ | ☐ | ☐ |

24. For each of the following statements, please choose the response that most represents your opinion or feeling:

| | Strongly Agree | Agree | Somewhat Agree | Neither Agree nor Disagree | Somewhat Disagree | Disagree | Strongly Disagree |
|---|---|---|---|---|---|---|---|
| a. I moved to [CITY] because I was dissatisfied with my home country. | ☐ | ☐ | ☐ | ☐ | ☐ | ☐ | ☐ |
| b. I moved to [CITY] because I was unhappy with the politics back home. | ☐ | ☐ | ☐ | ☐ | ☐ | ☐ | ☐ |
| c. My decision to move to [CITY] was mainly economic. | ☐ | ☐ | ☐ | ☐ | ☐ | ☐ | ☐ |
| d. I live here because property taxes are lower. | ☐ | ☐ | ☐ | ☐ | ☐ | ☐ | ☐ |
| e. I consider [CITY] to be my home. | ☐ | ☐ | ☐ | ☐ | ☐ | ☐ | ☐ |
| f. I feel at home in [CITY]. | ☐ | ☐ | ☐ | ☐ | ☐ | ☐ | ☐ |
| g. [CITY] feels more like a place where I live rather than a part of who I am. | ☐ | ☐ | ☐ | ☐ | ☐ | ☐ | ☐ |
| h. I am satisfied with my life here in [CITY]. | ☐ | ☐ | ☐ | ☐ | ☐ | ☐ | ☐ |
| i. I worry that too many Americans are ruining the [CITY]. | ☐ | ☐ | ☐ | ☐ | ☐ | ☐ | ☐ |
| j. I don't have a political voice living here. | ☐ | ☐ | ☐ | ☐ | ☐ | ☐ | ☐ |
| k. I live here because the health care is affordable. | ☐ | ☐ | ☐ | ☐ | ☐ | ☐ | ☐ |
| l. I live here because I can get better, cheaper care if I have a disability or develop Alzheimer's disease. | ☐ | ☐ | ☐ | ☐ | ☐ | ☐ | ☐ |
| m. I plan to continue to live in [CITY] even if I become disabled. | ☐ | ☐ | ☐ | ☐ | ☐ | ☐ | ☐ |
| n. I am interested in developing friendship with people who are part of the native community. | ☐ | ☐ | ☐ | ☐ | ☐ | ☐ | ☐ |
| o. I am interested in developing friendships with people who are part of the native community. | ☐ | ☐ | ☐ | ☐ | ☐ | ☐ | ☐ |
| p. I am more comfortable spending time with people from my own culture. | ☐ | ☐ | ☐ | ☐ | ☐ | ☐ | ☐ |
| q. I love to study and learn about local language and culture. | ☐ | ☐ | ☐ | ☐ | ☐ | ☐ | ☐ |
| r. I attend events that are put on by the local community and conducted in Spanish. | ☐ | ☐ | ☐ | ☐ | ☐ | ☐ | ☐ |
| s. I feel like I belong in [CITY]. | ☐ | ☐ | ☐ | ☐ | ☐ | ☐ | ☐ |
| t. I feel like the residents in [CITY] accept me. | ☐ | ☐ | ☐ | ☐ | ☐ | ☐ | ☐ |
| u. I have meaningful friendships with people who are part of the native community. | ☐ | ☐ | ☐ | ☐ | ☐ | ☐ | ☐ |

*We are almost finished.*

25. Please tell us about an experience you have had or an observation you've made that illustrates something that you consider important for others to know about living in [CITY].

Click or tap here to enter text.

26. In your opinion, what are the best things that a community can do to attract foreign retirees. Please list up to three things.

1.

2.

3.

27. In your opinion, what are three potential negative impacts of retirement migration on [CITY]?

1.

2.

3.

*About You*

28. What is your highest level of education?

☐  I did not complete high school
☐  High school graduate
☐  2-year college
☐  4-year college
☐  Post college

29. Please indicate your current employment status:

☐  Fully retired
☐  Part-time employment
☐  Full-time employment

30. What is your monthly income (in USD)?

☐  Less than $1,000 per month
☐  $1,000 - $3,000 per month
☐  Greater than $3,000 per month

31. With whom do you live? (you can answer more than one):

- ☐ Alone
- ☐ Spouse or Partner
- ☐ Other family member
- ☐ Friend
- ☐ Roommate (a person you house with, but are not otherwise connected to)
- ☐ Other: _____

32. Please choose the answer that best describes your current living situation in [CITY].

- ☐ Hotel
- ☐ Rental apartment or town house
- ☐ Apartment or town house I own
- ☐ House I rent
- ☐ House I own
- ☐ Other (please specify): _____

33. Do you live in a community that has controlled entry by a gate and/or guard?

- ☐ Yes
- ☐ No

34. Which one of the following best describes what you do with your cash savings (bank accounts, certificates of deposit):

- ☐ I have some cash or no cash in a bank in [CITY], but keep most of my cash savings in my home country
- ☐ I have accounts in both countries, and with approximately equal amounts in each
- ☐ I have most or all of my cash savings in one or more banks in [CITY]
- ☐ Other - please describe: _____

Thank you for taking the time to complete this survey! Your response will be a great help in further understanding retirement to foreign countries, and how this affects retirees as well as the cities themselves. If you are interested in obtaining a copy of the finished report from this study, please check the box and enter your email below.

If you have any comments on the survey or the project, feel free to leave a comment below as well!

- ☐ Yes, I would like to receive a copy of the report. Add your email: _____

Any additional comments? Feel free to leave them below:

# Index

© Springer Nature Switzerland AG 2020
P. D. Sloane et al. (eds.), *Retirement Migration from the U.S. to Latin American
Colonial Cities*, International Perspectives on Aging 27,
https://doi.org/10.1007/978-3-030-33543-4